Plant Alkaloids
A Guide to Their Discovery and Distribution

FOOD PRODUCTS PRESS
An Imprint of The Haworth Press, Inc.
Herbs, Spices, and Medicinal Plants
Lyle E. Craker, PhD
Senior Editor

Plant Alkaloids: A Guide to Their Discovery and Distribution
by Robert F. Raffauf

*Herbs, Spices, and Medicinal Plants: Recent Advances
in Botany, Horticulture, and Pharmacology,
Volumes 1-4,* edited by Lyle E. Craker and James E. Simon

Related titles of interest from Food Products Press:

Opium Poppy: Botany, Chemistry, and Pharmacology
by L. D. Kapoor

*The Honest Herbal: A Sensible Guide to the Use of Herbs
and Related Remedies, Third Edition* by Varro E. Tyler

Herbs of Choice: The Therapeutic Use of Phytomedicinals
by Varro E. Tyler

Plant Alkaloids
A Guide to Their Discovery and Distribution

Robert F. Raffauf, FLS, FAAAS

Food Products Press
An Imprint of The Haworth Press, Inc.
New York • London

Published by

Food Products Press, an imprint of The Haworth Press, Inc., 10 Alice Street, Binghamton, NY 13904-1580

Library of Congress Cataloging-in-Publication Data

Raffauf, Robert F. (Robert Francis), 1916-
    Plant alkaloids : a guide to their discovery and distribution / Robert F. Raffauf.
       p.   cm.
    Includes bibliographical references (p.   ) and index.
    ISBN 1-56022-860-1 (alk. paper)
    1. Alkaloids. 2. Botanical chemistry. I. Title.
QK898.A4R35   1996
581.19'242–dc20
                                                   96-5319
                                                    CIP

CONTENTS

ABOUT THE AUTHOR

Robert F. Raffauf, PhD, is Professor Emeritus of Pharmacognosy and Medicinal Chemistry at Northeastern University in Boston. He currently holds an appointment as Research Associate at the Botanical Museum of Harvard University, where he has also taught. He has served as Visiting Professor at the School of Pharmacy at the University of Puerto Rico and at the School of Biological Sciences of the National Polytechnic Institute of Mexico. Dr. Raffauf has led numerous expeditions in many parts of the world in the search for new plants of potential medicinal value and has lectured extensively on this and related matters, including the rain forests and conservation. He is the author of eight books, 69 journal publications, and four patents, and he continues to work with graduate students interested in natural products research. A 50-year member of the American Chemical Society, Dr. Raffauf is also a Fellow of the Linnean Society of London and of the American Association of Advancement of Science. In 1988, he was awarded an appointment as Resident Scholar at the Rockefeller Study Center in Bellagio, Italy.

Foreword

During several years of field work in the Northwest Amazonia, I lived and worked with members of many of the Amazonian Indian tribes. It was an extraordinary opportunity to observe, appreciate, and record their local customs, rituals, and particularly, as a botanist, their intelligent uses of the plants of the forests in which they lived. The importance of this information, beyond simply creating an interesting ethnobotanical record, was not entirely obvious at the time.

After my return to more academic pursuits, I met the author of the following pages, then a chemist for a major North American pharmaceutical company, with an interest primarily in that portion of my notes dealing with the treatment of disease as it was understood by the Indian peoples. Some plants, it was thought, could be sources of new chemical compounds for eventual use in our own system of medicine. Furthermore, in an attempt to reach that goal, the addition of chemical and pharmacological data to the botanical record would expand our knowledge of the rain forest and its plant and animal inhabitants. A collaboration seemed a natural and logical consequence.

As a result of the work described in this book, many plants have been the subjects of further botanical, chemical, and pharmacological research. This integrated, interdisciplinary approach has been of great advantage to our students, those in my courses at Harvard University as well as those of Professor Raffauf at Northeastern University. A number of our students have even carried out field work in various parts of tropical America. With them we have been able to co-author a number of technical papers on aspects of the wealth of natural resources in the Western Amazon and to supply them with challenging problems in the numerous disciplines bearing on rain forest science. Together we have published two books (*The Healing Forest, Vine of the Soul*) extending our knowledge of the biodiversity of this vast area of South America.

It is our hope that the information which my colleague has assembled here will continue to encourage academic as well as commercial research on the usefulness of plants to humans and contribute to current efforts at conservation of Amazonian resources, some of which are on the verge of extinction as a result of continued uncontrolled devastation in many areas of these marvelous forests.

–Richard Evans Schultes, PhD, FMLS
Jeffrey Professor of Biology,
Emeritus Director, Botanical Museum,
Harvard University,
Cambridge, Massachusetts

Preface

The search for plant alkaloids of novel chemical structure having potential value as medicinal agents, as toxic principles, or as appropriate starting materials for synthetic modification leading to other useful products, has occupied the attention of phytochemists for over 150 years. In 1950, about 2,000 of these substances were recognized; by 1970 this number had increased to about 4,000 and 20 years later 10,000 were known. In recent times, they have been considered as useful taxonomic markers in attempts to construct more "natural" systems of plant classification through chemotaxonomy, and as suitable substances for the study of biosynthetic pathways in plant metabolism. During the latter part of the present century, emphasis on the conservation of plant resources and the ethnobotanical information concerning their use by many of the world's aboriginal societies has given added impetus to the importance of the continued study of the "chemical factory," represented by the large unexploited portion of the plant kingdom before much of it disappears under the pressures incident to the mass movements of peoples and the increase in the world's population. Both are in large part responsible for devastation of many floras, particularly those of the rain forests. Not only will a number of species be lost even before they are known and named by botanists, but literally thousands of chemical compounds new to science will disappear forever.

During the past 40 years I have been involved, in one way or another, with the screening of several thousand plants for the presence of alkaloids as potential medicinal agents under the auspices of a number of governmental, industrial, and academic institutions. Underlying this activity has been the hope that the discovery of new compounds of this class would lead to substances at least as useful as those which similar studies have produced in the past. This screening has been done on fresh plant material in the field, on small quantities taken from herbarium specimens, and in the laboratory using a few grams of dried material made available by botanists, collectors, herb dealers, and my personal collections in many parts of the world.

Various methods for the screening of large numbers of plant samples for alkaloids have been used by many investigators more or less successfully (Farnsworth, 1966). Several of these were used, depending on the facilities available at the time, but most of the results reported herein were obtained by simple methods described some years ago (Raffauf, 1962a; Raffauf and Altschul, 1968). In the field, these involve spotting a droplet of plant sap on filter paper and applying a droplet of Dragendorff's reagent; the development of a red-orange color indicates the presence of alkaloids. In the laboratory or herbarium, simple extracts of dried plant material may be used, with certain limitations, for the same purpose (Balick, Rivier, and Plowman, 1982).

Methods may be adapted to needs of the investigator; tests for some specific types of nitrogenous compounds may be included (e.g., indoles, simple amines, and amino acids). An approximation of the quantity of alkaloid in a sample may be made by comparing the intensity of the color produced in the Dragendorff test with those produced by standard alkaloid solutions of known concentration. By using the Dragendorff reagent as a spray, it is also possible to conduct thin-layer chromatographic studies in the field. Several years ago, in an attempt to devise a method for the identification of specific compounds in a particular alkaloid-positive collection, a portable laboratory was assembled for the evaluation of small extracts of fresh plant material by chromatographic analysis using alumina-coated microscope slides and samples of the alkaloids expected to be present. In the course of the study, it was found that some of the compounds were present only during a restricted portion of the plant's growth cycle. We now know that, in some cases at least, alkaloids are indeed further modified by the plants that produce them.

An advantage of these simple methods is that they allow such studies to be done far from a source of electric power and other amenities of the laboratory. But it is also true that there are a number of uncertainties in such procedures; not all nitrogen-containing substances will react with either Dragendorff's or Mayer's reagent. A battery of test reagents would give a more definitive although, even then, not an infallible result (Abisch and Reichstein, 1960). False positive tests are given by many types of nonalkaloidal plant constituents with a variety of alkaloidal reagents (Habib, 1980). Balick, Rivier, and Plowman (1982) have pointed out the importance of methods used in field drying and preservation of herbarium specimens with respect to the reliability of the results obtained when testing them. A plea for such testing and a review of the more elegant methods for its accomplishment has been given by Philippson (1982). Furthermore, as every plant collector has discovered, it is not always practical in a given instance to collect all of the parts of a plant in which alkaloids may occur. Nonetheless, an estimated 85 percent of alkaloid-containing plants can be detected by the methods described here; a number of known alkaloidal plants have been included in the survey to serve as controls.

Herbarium specimens representing otherwise relatively inaccessible species of several families (Apocynoceae, Bombacaceae, Lycopodiaceae, Lythraceae, Orchidaceae, Rubiaceae) were included in this survey. Small samples were selected from sheets in the Gray, Oakes Ames, and Arnold Arboretum herbaria of Harvard University under the guidance of Professor Richard Evans Schultes, Emeritus Director of the Botanical Museum, whose assistance is gratefully acknowledged.

Not all of this testing was done by me; some of it was done by anthropologists, ethnobotanists, and plant collectors in the course of field work sponsored by academic or industrial programs under my direction, some by laboratory technicians under my supervision, some by former students as preliminary exercises in phytochemistry, and some in collaboration with phytochemical programs supported by the Councils of Scientific and Industrial Research of Australia and South Africa. Portions of the test results from these programs which had been at my disposal, have been included here for completeness in order to convey some idea of the alkaloid distribution in plant families represented in the southern hemisphere. Further data on these studies, as well as the results of the isolation and pharmacological testing of a large number of alkaloids, are to be found in a recent excellent publication by the Melbourne group (Collins et al., 1990).

With these reservations, the ensuing compilation is offered as a guide to the distribution and discovery of potentially new alkaloids in species from over 300 plant families. It summarizes the result of tests on about 30,000 samples representing some 19,000 species in about 4,000 genera. Doubtful or "trace" tests were not recorded; samples identified only to family at the time of collection and assay have not been included. In many instances, several samples of the same species were tested in recognition of the fact that alkaloid content often varies with the extent of plant growth, season, soil, climate, and, in some cases, with the time of day at which the collection was made. In listing the test results, positives representing known alkaloidal plants are given first, followed by a list of the new species and these in turn by those which were negative. The positive tests are recorded as a fraction–the number obtained over the number of samples tested (if more than one).

Through the years in which this survey was carried out, taxonomists have sometimes shifted genera from one family to another or created new genera and even new families in an effort to refine earlier systems of classification and to accommodate new plant discoveries. With a few exceptions, the family assignments used here follow Mabberley [Cronquist] (1989). The genera and species cited are those given by the collectors or suppliers and no attempt to correct or modify these assignments has been made other than to validate the generic names and their synonymies as given by Mabberley (1989) and Willis (1985).

The chemical references cited in the bibliography are standard compilations devoted to the more than 10,000 known alkaloids and should be familiar to those scientists with interest in these compounds. References beyond those have been added, where appropriate, at the end of the plant families discussed. References to less common journals are accompanied by notation of the Chemical Abstracts (Chem. Abs.) citation of the original article.

Unfortunately, I am no longer able to supply collectors' numbers for all voucher specimens that were prepared in the course of these studies. But in most instances, except for a few samples obtained from herb dealers and local markets, such specimens were deposited in the national or university herbaria in the geographical area of origin and in the Economic Botany Herbarium of the Botanical

Museum of Harvard University. Using the standard abbreviations for the herbaria of the world, these include the institutions listed in Appendix A.

The plant parts tested were usually leaves and stems. When ethnobotanical data suggested the importance of other plant parts (e.g., roots, bark) these were mentioned along with the results obtained. Of the approximately 19,000 species tested, positive indications for the presence of alkaloids were obtained in about 3,600 (19 percent) of which 3,200 were new, based on literature available at the time of preparation of this manuscript. These were found in a total of 315 families: 48 of gymnosperms and ferns, 43 of monocots, and 224 of dicots including 134, 199, and 2,900 species respectively. However, because this survey was not entirely random–many well-known alkaloidal genera/species were avoided except for the inclusion of a few from time to time to serve as controls for the test procedures– and because of the uncertainties inherent in the methods that were mentioned earlier, these figures do no more than suggest alkaloid distribution throughout the plant kingdom. Nonetheless, this information should be sufficient for the interested phytochemist to undertake the discovery of new and potentially useful compounds.

In one way or another, literally hundreds of people have contributed to this endeavor, from individuals in the native populations of New Guinea, Africa, and the Amazon, to ministers in the governments on all continents except Antarctica. Especially, I acknowledge the early encouragement of Dr. Glenn E. Ullyot and the staff at Smith Kline & French Laboratories during the 1950s and 1960s and for permission to include results of testing done under their auspices. To these and to the botanists, ethnologists, anthropologists, industrialists, missionaries, teachers, students, and interested laypersons who have given their assistance, this compilation is gratefully dedicated. May it lead, if only in a small way, to an increased awareness of the need for the continued study of the environment, the conservation and prudent use of the natural resources on which we all depend, and eventually to the discovery of a few more useful agents in our constant battle against hunger and disease.

–R.F.R.
Boston, MA

ALKALOID TEST RESULTS

A

ACANTHACEAE
346 genera; 4,300 species

This is a pantropical family with four centers of diversification: Amazon, Central America, Africa, and Indo-Malaysia. Its classification has not been, and may not yet be, a matter of agreement among taxonomists, but at the moment the family would seem to be divided into three subfamilies with a close relationship of several members to the Scrophulariaceae. Some members are cultivated as ornamentals.

Alkaloids have been detected previously in a few genera. In this study of about 400 samples representing 297 species, six known alkaloidal species were included: *Acanthus ilicifolius* (1/2), *Adhatoda vasica*, *Anisotes sessiliflorus*, *Hypoestes verticillaris* (2/4), *Macrorungia longistrobus*, *Rhinacanthus communis* (1/2).

Alkaloids were also detected in the following: *Angkalanthus transvaalensis* (2/2), *Anisacanthus insignis*, *Aphelandra deppeana* (1/4), *Asystasia atriplicifolia*, *A. welwitchii*, *Barleria matopoensis*, *B. rotundifolia*, *B. sinensis*, *Blechum pyramidatum* (1/3), *Blepharis boerhaavifolia*, *B. marginata*, *B. natalensis*, *Blepharis sp.*, *Crossandra spinescens*, *Dicliptera clinipodia*, *Duvernoya* (= *Justicia*) *aconitifolia*, *D. adhatoides* (1/2), *Dyschoriste hirsutissima* (1/4), *Ecbolium sp.*, *Elytraria acaulis*, *E. squamosa* (3/3), *Hemigraphis hirta* (whole plant), *Hemigraphis spp.* (2/2), *Hypoestes aristata* (1/3), *Jacobinia* (= *Justicia*) *spicigera* (1/2), *Justicia americana*, *J. anselliana*, *J. elegantula*, *J. flava* (1/2), *J. montana*, *J. orchioides*, *J. protracta* (2/2), *J. salviaefolia*, *J. thymifolia*, *J. trinervia*, *J. ventricosa*, *Mirandea grisea*, *Monechma atherstonei*, *M. australis*, *M. incanum*, *Monechma sp.* (1/3), *Neuracanthus africanus* (1/2), *Orthotactus montanus* (1/2, leaves and flowers), *Peristrophe cernua*, *Phlogacanthus thyrsiflorus* (1/2, root), *Ruttya ovata* (1/2), *Sanchezia thinophila* (bark), *Siphonoglossa ramosa*.

3

The alkaloids of *Adhatoda vasica* have been reviewed (Jain, 1984). *Acanthus ilicifolius* contains benzoxazoline-2-one; the alkaloids of *Acanthus mollis* have been reported (Wolf et al., 1985) and new spermine-type alkaloids have been isolated from *Aphelandra pilosa* (Tawil et al., 1989). In view of the native use of at least one species of *Justicia* as a hallucinogen in South America (Schultes and Holmstedt, 1968), the report of its presumed content of tryptamine needs corroboration. In this connection, the several listed alkaloid-positive species from other parts of the world should be of interest.

Negative tests were obtained from the following species: *Acanthopsis carduifolia, Acanthus ebracteatus, A. mollis, Adenosma glutinosum, Adhatoda sp., Ancylacanthus bainesii, Anisacanthus gonzalezii, A. quadrifolius, A. ochoterenae, A. thurberi, A. tulensis, A. wrightii, Anisotes formosissimus, Aphelandra auriantiaca, A. blanchetiana, A. chamissoniana, A. deppeana, A. incerta, A. pilosa, Alphelandra sp., Asteracantha (= Hygrophila) spinosa, Asystasia gangetica, A. schimperi, A. varia, Barleria albostellata, B. cristata, B. crossandriformis, B. discolor, B. elegans, B. guensii, B. heterotricha, B. kirkii, B. lancifolia, B. lugardii, B. micans, B. obtusa, B. pretoriensis, B. prionitoides, B. pungens, B. pyramidata, B. randii, B. rigida, B. scandens, Barleria sp., Beloperone (= Justicia) californica, B. comosa, B. fragilis, B. guttata, Beloperone sp., Blechum nipponicum, B. plagiogyriflorus, Blechum sp., Blepharis capensis, B. diversispina, B. glumacea, B. maderaspatensis, Blepharis sp., B. squarrosa, Bravaisia integerrima, Carlowrightia glabrata, C. glandulosa, C. serpyllifolia, Carlowrightia spp. (2), Chaetacanthus setiger, Chaetothylax hatschbachii, Chileranthemum violaceum, Codonacanthus pauciflorus, Crabbea angustifolia, C. hirsuta, Crossandra greenstockii, C. undululaefolia, Cyrtanthera pohliana, Daedalacanthus (= Eranthemum) montanus, D. nervosus, D. purpurescens, Dianthera (= Justicia) ovata, Diapedium (= Dicliptera) assurgens, D. chinensis, D. micranthus, D. nobilis, D. peduncularis, D. pringlei, D. resupinata, D. rigidissima, Disperma (= Duosperma) crenatum, Dyschoriste decumbens, D. fischeri, D. ovata, D. microphylla, D. quadrangularis, D. rogersii, Dyschoriste sp., D. verticillaris, Ebermaiera (= Staurogyne) corniculata, Ecbolium amplexicaule, E. linnaeatum, E. revolutum, Elytraria bromoides, Eranthemum eldorado, E. nervosum, Graptophyllum pictum, Grap-*

tophyllum sp., *Haplanthus nilgherriensis*, *Hemigraphis elegans* var. *crenata*, *H. hirta*, *H. latebrosa*, *Hemigraphis spp.* (2), *Henrya* (= *Tetramerium*) *yucatanensis*, *Hygrophila laxifolia*, *H. salicifolia*, *Hygrophila spp.* (4), *H. spinosa*, *Hypoestes floribunda*, *H. phalopsoides*, *H. purpurea*, *Isoglossa grantii*, *Isoglossa sp.*, *I. stipitata*, *I. woodii*, *Jacobinia* (= *Justicia*) *aschenborniana*, *J. candicans*, *J. heterophylla*, *J. incana*, *J. mexicana*, *J. paniculata*, *J. sellowiana*, *Jacobinia spp.* (2), *J. stellata*, *Justicia angalloides*, *J. betonica*, *J. betonicoides*, *J. beyrichii*, *J. brasiliana*, *J. campechiana*, *J. campylostemon*, *J. cheirianthifolia*, *J. furcata*, *J. gendarussa*, *J. kirkiana*, *J. kraussii*, *J. mexicana*, *J. odorata*, *J. ovata*, *J. petiolaris*, *J. procumbens*, *J. secunda*, *Justicia spp.* (5), *Lepidagathis formosensis*, *L. incurva*, *L. microchila*, *L. persimilis*, *Lepidagathis sp.*, *Mackaya bella*, *Macrorungia formosissima*, *Mendoncia coccinea*, *M. hoffmannseggiana*, *M. sellowiana*, *Mendoncia sp.* (this genus is sometimes placed in a family of its own, Mendonciaceae), *Monechma debile*, *M. divaricatum*, *M. fimbricatum*, *M. molissium*, *M. pseudopatulum*, *M. scabridum*, *Monechma spp.* (2), *Odontonema calystachum*, *O. cuspidatum*, *Odontonema spp.* (2), *Pachystachys coccinea*, *Peristrophe bicalyculata*, *P. grandibrachiata*, *P. natalensis*, *Petalidium aromaticum*, *P. barlerioides*, *P. bracteatum*, *P. oblongifolium*, *P. rubescens*, *Phaulopsis betonica*, *P. imbricata*, *Phlogacanthus thyrsiflorus*, *Pseuderanthemum praecox*, *Pseuderanthemum spp.* (2), *Rhinacanthus xerophilus*, *Ruellia alba*, *R. albicaulis*, *R. albiflora*, *R. bourgei*, *R. colorata*, *R. cordata*, *R. formosa*, *R. inundata*, *R. macrophylla*, *R. nudiflora*, *R. nudiflora* var. *yucatana*, *R. palmeri*, *R. patula*, *R. peninsularis*, *R. pilosa*, *R. prostrata*, *Ruellia spp.* (4), *R. speciosa*, *R. tuberosa*, *R. tweediana*, *Rungia parviflora*, *Ruspolia hypocrateriformis*, *Sanchezia nobilis*, *Sclerochiton harveyanus*, *Sericographis* (= *Justicia*) *cordifolia*, *Siphonoglossa pilosella*, *S. tubulosa*, *Stenandrium barbatum*, *Strobilanthes cusia*, *S. formosanus*, *Teliostachya* (= *Lepidagathis*) *alopecuroides*, *Tetramerium aureum*, *T. hispidum*, *T. sureum*, *Thunbergia amoena*, *T. atriplicifolia*, *T. erecta*, *T. fragrans*, *T. grandiflora*, *T. lancifolia*, *T. natalensis*, *Thunbergia sp.* (the genus is sometimes placed in Thunbergiaceae), *Thyrsacanthus* (= *Odontonema*) *callistachyus*, *Tricanthera gigantea*.

REFERENCES

Jain, M. P., *Journal of Indian Drugs 21* (1984) p. 313.

Schultes, R. E. and B. Holmstedt, *Rhodora 70* (1968) p. 113.

Tawil, B. F., J. Zhu, U. Prantini, and M. Hesse, *Helvetica Chimica Acta 72* (1989) p. 180.

Wolf, R. B., G. F. Spencer, and R. D. Prattner, *Journal of Natural Products 48* (1985) p. 59.

ACERACEAE
2 genera; 113 species

This is a north temperate family with some distribution in the mountains of the tropical zone. Trees are valued as ornamentals, as lumber, and, in northern North America at least, as the basis for the maple sugar industry (*Acer saccharum*). Of the species in the family, all but ten are assigned to the genus *Acer*.

The discovery of gramine in the leaves of *Acer rubrum* and *A. saccharinum* prompted an examination of other members of the genus, five of which had been reported to give positive tests for alkaloids. One hundred and sixty-two samples representing 69 species and their varieties were tested; positive results were obtained with *A. rubrum* (1/4), *A. rubrum* var. *pycnanthum* and *A. saccharinum* (2/6). It would be of interest to discover the role of gramine in the economy of the important species of *Acer* in which it is found. The toxic amino acids hypoglycin A and B have been found in the seeds of *A. pseudoplatanus*.

In this study, the following were alkaloid-negative: *Acer barbatum, A. buergerianum, A. campestre* var. *leiocarpum, A. cissifolium, A. crataegifolium, A. diabolicum, A. dieckii, A. duretti, A. formosanum, A. ginnale, A. ginnale* var. *durand, A. griesium, A. glabrum, A. heldreichii, A. heldreichii* var. *macropterum, A. japonicum* var. *aconitifolium, A. kawakamii, A. leucoderma* var. *saccharum, A. macrophyllum, A. mandshuricum, A. mayrii, A. miyabei, A. monosepalum, A. negundo, A. negundo* var. *nanum, A. negundo* var. *violaceum, A. nikoense, A. palmatum* var. *dissectum,* var. *heptalobum,* var. *multifidum, A. pennsylvanicum, A. platanoides, A. platanoides* var. *cleveland,* var. *globosum,* var. *harlequin,* var. *natorp,* var. *palmatifidum,* var. *rubrum,* var. *undulatum,* var. *variegatum,*

A. pseudoplatanus, var. *nizetti, A. pseudosieboldianum, A. rubrum, A. rubrum* var. *columnare*, var. *scanlon*, var. *schlesinger*, var. *tilford, A. rufinerve, A. saccharum* var. *pyramidale*, var. *lacinatum*, var. *glaucum*, var. *golden*, var. *lutescens*, var. *tripartitum, A. serratum, A. sieboldianum, Acer sp., A. spicatum, A. tartaricum, A. tegmentosum, A. triflorum, A. truncatum, A. tschonoskii, A. zoeschense.*

ACHARIACEAE
3 genera; 3 species

This is an African family with a single species in each of three genera. It is related to the Passifloraceae and the Cucurbitaceae. The only sample tested, *Ceratiosicyos ecklonii*, was alkaloid-negative.

ACITINDIACEAE
3 genera; 355 species

The family is distributed throughout eastern Asia and south to northern Australia with some representatives in tropical America. It is related to the Dilleniaceae. The familiar kiwi fruit, *Actinidia chinensis*, is a member of this family as are a few others cultivated as ornamentals.

Two species had been reported in earlier literature as alkaloid-positive, but *Actinidia arguta, A. chinensis, Saurauia oldhamii, A. pseudorubiformis, Saurauia spp.* (5), and *S. villosa* gave negative tests in the present study.

AGAVACEAE
18-20 genera; 410-670 species

Previously included, at least in part, in the Liliaceae or Amaryllidaceae, the Agavaceae, familiar plants in arid America, are known for their content of steroidal saponins which have been used for conversion to bioactive steroids of medicinal importance. It is otherwise known as the source of fibers such as istle, sisal, and others.

The presence of alkaloids had been reported in a species of *Nolina*, in a few species of *Dracaena*, in *Cordyline terminalis* (tyramine), in *Yucca whipplei*, and in an undetermined species of *Manfreda*.

In this study, positive alkaloid tests were obtained for *Dasylirion acrotrichum* (1/3), *D. cedrosanum* (1/2), *Dracaena steudneri*, *D. usumbarensis*, *Phormium colensoi*, *Sansevieria aethiopica*, *S. deserti* (1/2), *S. grandis*, *S. thyrsiflora* (1/3), *Yucca carnerosana*, *Y. decipiens*, *Y. elephantippe*.

Negative tests were obtained for the following: *Cordyline australis*, *C. baueri*, *C. dracaenoides*, *C. fruticosa*, *C. indivisa*, *Cordyline spp.* (3), *Dasylirion palaciosii*, *Dracaena fragrans*, *D. angustifolia*, *D. hookeriana*, *Dracaena sp.*, *Nolina bigelovii*, *N. interrata*, *N. parviflora*, *N. texana*, *Phormium tenax*, *Yucca aloifolia*, *Y. brevifolia*, *Y. filifera*, *Y. rigida*, *Y. smalliana*, *Yucca sp.*, *Y. treculeana*, *Y. whipplei*.

AIZOACEAE
Over 100 genera; up to 2,400 species

This is primarily a South African family but representation exists in tropical Africa, Asia, Australia, South America, and the United States. It is important as a source of ornamentals. In South Africa, several species have been suggested as hallucinogens.

The number of genera and their classification in the family is not yet a matter of agreement among taxonomists. The major genus (anywhere from 300 to 800 species, and sometimes placed in a family of its own, Mesembryaceae) is *Mesembryanthemum*, which yields the alkaloid mesembrine and its relatives. Alkaloids have been reported in a few other genera; they may be precursors of the betacyanins and betaxanthins found in several members of the family. In view of the confusing state of the taxonomy of the family, further work on the utility of the alkaloids as taxonomic markers seems indicated.

In the earlier literature there are reports of the detection in or isolation of alkaloids from 46 species, 24 of which had been included in *Mesembryanthemum*. Some of these are now recognized

as distinct, *viz*: *Aptenia, Drosanthemum, Glottiphyllum, Laparanthus, Mestoklema, Aridaria, Oscularia, Prenia, Ruschia, Trichodiadema.*
The following species, known to be alkaloid-positive from earlier work, were also found to be positive in the survey reported here: *Psilocaulon absimile* (2/2), *Sceletium joubertii, S. tortuosum.*
Positive tests were likewise obtained for the following: *Delosperma tradescantioides* (2/2), *Galenia africana* (2/3), *G. fruticosa, G. procumbens, G. sarcophylla, Hypertelis salsoloides* (2/2), *Mestoklema arboriforme, M. tuberosum, Mollugo mulligenes, Pharnaceum brevicaule, Plinthus sericeus, Psilocaulon granulicaule, Ruschia griquesne, Sceletium anatanicum, S. rigidum, Zaleya pentandra.*
Negative tests were obtained for the following species: *Aizoon zygophylloides, Aptenia cordifolia, Brownanthus ciliatus, Carpobrotus edulis, C. rossei, Cephalophyllum alstonii, Cephalophyllum sp., Chasmatophyllum musculinum, Conicosia sp., Conophytum odoratum, Corbichonia decumbens, Delosperma asperulum, D. cooperi, D. hirtum, D. pageanum, Delosperma spp.* (2), *Drosanthemum floribundum, D. haworthiae, D. jamesii, D. speciosum, Erepsia includens, Gisekia africana, Glinus bainesii, G. lotoides, G. radiatus, Hypertelis bowkeriana, Lampranthus recurvus, Lampranthus spp.* (2), *Macarthuria neocambrica, Malephora thunbergii, Mesembryanthemum aequilaterale, M. australe, M. crystallium, Mollugo cerviana, M. pentaphylla, M. verticillata, Pharnaceum aurantium, Plinthus karrodicus, Prenia pallens, Psilocaulon spp.* (2), *Ruschia festiva, R. hamata, R. indurata, R. multiflora, R. ringens, Ruschia sp., R. uncinella, Sesuvium maritimum, S. portalacastrum, Sesuvium sp., Tetragonia arbuscula, T. decumbens, T. expansa, T. fruticosa, T. implexicoma, T. portulacoides, T. spicata, T. tetragonoides, T. verrucosa* (Tetragonia is sometimes put in a family of its own, Tetragoniaceae), *Trianthema decandra, T. erectum, T. pilosa, T. portulacastrum, T. transvaalensis, Trichodiadema pomeridianum.* The following have been placed in Molluginaceae: *Corbichonia, Glinus, Hypertelis, Mollugo, Pharnaceum.*
Alkaloids have been identified in *Aizoon canariense, Mesembryanthemum forskahlei,* and *N. nodiflorum* (Rizk et al., 1986). Several new alkaloids have been isolated from *Sceletium* (Jeffs, Capps, and

Redfearn, 1981). Their biosynthesis of the *Sceletium* alkaloids has been studied (Herbert and Kattah, 1989).

REFERENCES

Herbert, R. B. and E. Kattah, *Tetrahedron Letters 30* (1989) p. 141.
Jeffs, P. W. in *The Alkaloids 19* (1981) p. 1, Academic Press, New York.
Jeffs, P. W., T. M. Capps, and R. Redfearn, *Journal of Organic Chemistry 47* (1982) p. 3611.
Rizk, A.M., H. I. Heiba, H. A. Ma'ayevgi, and K. H. Batanouny, *Fitoterapia 57* (1986) p. 1.

ALANGIACEAE
1 genus; 17 species

The one genus, *Alangium*, of the tropics and semitropics of the Old World, is rich in alkaloids, not all of which have had structural assignments. Considerable synonymy exists in the family. The chemistry of *Alangium lamarckii* has been studied in some detail and positive tests for alkaloids were obtained from the single sample of Indian origin included in this study.

ALISMATACEAE
11 genera; 95 species

This is a cosmopolitan family but it occurs mainly in temperate and tropical regions of the northern hemisphere. Some species are used as ornamentals, others are familiar aquarium plants, and the roots of *Sagittaria* are used as food in China. Twenty-two samples of 16 species were tested and positive results were given by *Echinodorus radicans* (2/2), *Sagittaria engelmanniana, S. graminea* (1/2), and *S. latifolia*. The family is not known for the presence of alkaloids; there are but two earlier reports of their occurrence.

Negative tests were obtained for *Alisma plantago, A. plantagoaquatica, A. subcordatum, A. triviale, Caldesia parnassifolia, Echinodorus cordifolius, E. grandiflorus, E. virgatus, Limophyton obtu-*

sifolium, *Lophotocarpus* (= *Sagittaria*) *guayamensis*, *Sagittaria lancifolia*, and *S. sagittarifolia*.

ALSTROEMERIACEAE
4 genera; 200 species

This small group of Central and South American plants has been considered by some taxonomists as a family in its own right. Others have placed it as a division of the Amaryllidaceae. Mabberley, who follows Cronquist's system of classification, now lists even the Amaryllidaceae as a subdivision of the Liliaceae. The chemistry of these taxa is sufficiently different to argue for their separate family status, which will be maintained here.

No alkaloids are known nor were they detected in 15 samples representing four species of *Alstroemeria* and nine of *Bomarea*: *Alstroemeria inodora*, *A. pelegrina*, *Alstroemeria spp.* (2), *Bomarea acutifolia*, *B. edulis*, *B. hirtella*, *B. orata*, *B. salicoides*, *Bomarea spp.* (4).

ALOEACEAE
7 genera; 400 species

The family, characteristic of Arabia and South Africa with some species in other parts of Africa and Madagascar, has been separated from the Liliaceae. Species have been introduced elsewhere. Several have been used as a source of laxative anthraquinones and as a component of cosmetic preparations. Aloe is one of the oldest drugs.

Positive alkaloid tests are apt to be due to the formation of complexes of nonalkaloidal constituents with the Dragendorff reagent; alkaloids are not known in the family.

Positive tests were obtained here with *Aloe camronii*, *A. decurva*, *A. excelsa*, *A. globuligemma*, *A. littoralis*, *A. munchii*, *A. ortholopha* (2/2), and *A. suffulta*.

On the other hand, 17 other species of *Aloe*, one of *Gasteria*, and three of *Haworthia* were negative: *Aloe chabaudii*, *A. christianii*,

A. ciliaris, A. claviflora, A. cryptopoda, A. dichotoma, A. ecklonis, A. garipensis, A. greenii, A. hereroensis, A. melsetterensis, A. microstigma, A. plicatilis, A. saponaria, A. tenuior, A. vera, A. zebrina, Gasteria sp., Haworthia fouchei, H. margaritifera, H. setara.

AMARANTHACEAE
71 genera; 800 species

The family is most abundant in tropical America and Africa, but 18 of the 20 New World genera are found in the United States. A few genera are used for food, a few as ornamentals, and several are considered weeds.

Nitrogenous compounds are known (betacyanins and betaxanthins) but no alkaloids in the strictest sense have been described. Betaines often give positive tests for alkaloids, which may account for the report of several "unnamed alkaloids" in nine genera.

The following species gave positive tests: *Achyranthes aspera* (1/10) (previously known), *Alternanthera mexicana, A. repens, Amaranthus albus* (1/4), *A. graecizans* (4/4) (previously known), *Celosia trigyna* (4/11), *Centemopsis gracilenta, Cyphocarpa angustifolia, Iresine discolor, I. schaffneri* (1/3), *Iresine sp., Nothoserva brachiata, Pfaffia iresinoides* (2/2), *Pupalia caparia* (1/2), *Sericocoma remotiflora, S. sericea, Tidestromia lenuginosa* (1/2).

Negative tests were obtained for the following: *Achyranthes argentea, A. bidentata, A. indica, A. longifolia, A. ramosissima, A. rubrofusca, Achyranthes sp., Achyropsis laniceps, A. leptostachya, Acnida* (= *Amaranthus*) *cannabina, A. tamaracina, Aerva leucura, Alternanthera achyrantha, A. lehmannii, A. nodiflora, A. philoxeroides, A. pungens, A. sessilis, A. versicolor, Amaranthus alnus, A. australis, A. caudatus, A. hybridus, A. interruptus, A. palmeri, A. paniculatus, A. retroflexus, A. rubra, Amaranthus spp. (4), A. spinosus, A. viridis, Brayulinea* (= *Quilleminea*) *densa, Celosia argentea, C. linearis, C. nitida, Centema subfusca, Chamissoa sp., Cyathula crispa, C. cylindrica, C. oncinulata, C. orthocantha, C. prostrata, Cyphocarpa angustifolia, Deeringia arborescens, D. celosioides, D. polysperma, Dicraurus leptocladus, Digera arvensis, Froelichia floridana, F. gracilis, Gomphrena agrestis, G. canescens, G. celosioides, G. decumbens, G. graminea, G. gna-*

phaloides, G. incana, G. macrocephala, G. scapigena, G. nitida, G. sonorae, Gomphrena spp. (3), *Hermbstaedtia odorata, Iresine canescens, I. cassinaeformis, I. celosia, I. celosioides, I. grandis, I. heterophylla, I. interrupta, Pandiaka lindiensis, P. schweinfurthii, Pandiaka sp., Pfaffia glauca, P. paniculata, P. pulverulenta, Philox-erus portulacoides, Pseuderanthemum reticulatum, Ptilotus alope-curoides, P. atriplicifolia, P. helipteroides, P. nobilis, P. obovatus, P. polystachys, P. spicatus, Pupalia coparia, P. lappacea, Sericore-ma remotiflora, S. sericea, Telanthera polygonoides, Tidestromia (= Alternanthera) oblongifolia.*

AMARYLLIDACEAE
85 genera; 1,100 species

Long recognized as a family of their own, the amaryllids are now included in the Liliaceae by Cronquist but perhaps not generally. They are distributed throughout the world, mostly in the tropics and subtropics, and valued for their garden flowers (*Amaryllis, Crinum, Lycoris, Narcissus,* etc.).

Many amaryllidaceous alkaloids are known; at least 190 species and their horticultural varieties in over 30 genera have been reported to contain them. A large number have been characterized chemically, and they are sufficiently distinct from those in the Liliaceae to suggest separate family status, which will be maintained here.

Because of what appeared to be a lack of "useful pharmacological activity" associated with the amaryllids, little emphasis was placed on attempts to discover new ones in this survey. A review of the alkaloids is available (Grundon, 1989).

Forty-five samples representing 35 species were tested; nine had been known to contain alkaloids: *Amaryllis belladonna, Ammocharis coranica, Crinum asiaticum, C. giganteum, Haemanthus multiflorus, Hippeastrum vittatum, Lycoris radiata, Narcissus pseudonarcissus, Sprekelia formosissima.*

Positive tests were also obtained for the following: *Amaryllis sp., Clivia caulescens, Crinum brisbanicum* (1/2), *C. bulbospermum, C. commelynii, C. defixum, C. kunthianum, C. redunculatum, Cybistetes longifolia, Haemanthus magnificus, H. rotundifolius,*

Hippeastrum puniceum, Hippeastrum sp., Hymenocallis keyensis, H. occidentalis, Narcissus sp., Zephyranthes atamasco, Z. carinata, Z. rosea.

Negative results were obtained for *Amaryllis vittatum, Anoiganthus* (= *Cyrtanthus*) *breviflorus, Bravoa geminiflora, Brodiaea pulchella, Crinum macrowanii, C. macrantherum,* and an unidentified *Crinum* species.

REFERENCE

Grundon, M. F., *Natural Products Reports* 6 (1989) p. 79.

ANACARDIACEAE
73 genera; 850 species

Representatives of this mainly tropical family extend into north temperate regions of Eurasia and North America. Several are of economic importance (cashew and pistachio nuts, mango fruit, lacquer) and some are known for their content of substances highly irritating to the skin (urushiol and its relatives). Seven species have been reported to give positive tests for alkaloids. In this study, 170 samples from 111 species gave only one positive test for a species previously considered alkaloidal, *Dracontomelon magniferum.* Others found positive in this survey included *Astronium flaxinifolium* (2/2), *Buchanania arborescens, Lannea stuhlmannii, L. welwitchii, Rhus angustifolia, R. ciliata, R. incisa,* and *R. virens* (1/2).

The following species were negative: *Actinocheita filicinia, Anacardium giganteum, A. microcephalum, A. occidentale, Astronium graveolens, A. microcalyx, Astronium sp., A. ulei, Blepharocarya involucrigera* (in Blepharocaryaceae by some authorities), *Buchanania heterophylla, Comocladia platyphylla, Comocladia sp., Cotinus coggygria, Dobinea vulgaris, Dracontomelon dao, D. sylvestre, Euroschinus papuanus, Harpephyllum caffrum, Heeria argentea, H. dispar, H. insignis, H. paniculosa, H. reticulata, Heeria spp.* (4), *H. stenophylla, Lannea discolor, L. edulis, Laurophyllus capensis, Lithraea brasiliensis, L. molleoides, Mangifera indica, Odina* (= *Lannaea*) *wodier, Ozoroa reticulata, Pistacia chinensis, P. integerrima, P. mexicana, Protorhus longifolia, Rhodosphaera rhodan-*

thera, *Rhus amerina*, *R. batophylla*, *R. copallina*, *R. dentata*, *R. dissecta*, *R. dregeana*, *R. dura*, *R. ernesti*, *R. erosa*, *R. glabra*, *R. glauca*, *R. hypoleuca*, *R. integrifolia*, *R. intermedia*, *R. kirkii*, *R. lancea*, *R. laurina*, *R. legati*, *R. leucantha*, *R. longipes*, *R. longispina*, *R. lucida*, *R. magalismontana*, *R. microphylla*, *R. mollis*, *R. natalensis*, *R. ovata*, *R. pyroides*, *R. quartiniana*, *R. rehmanniana*, *R. rigida*, *R. rosmarinifolia*, *R. simii*, *Rhus spp.* (2), *R. spinescens*, *R. succedema*, *R. traitensis*, *R. tenuinervus*, *R. terebinthifolia*, *R. tomentosa*, *R. trilobata*, *R. typhina*, *R. undulata*, *Schinus engleri*, *S. molle*, *Schinus sp.*, *S. terebinthifolius*, *S. weimanniifolius*, *Sclerocarya caffra*, *Semecarpus atra*, *S. cuneiformis*, *Smodingium argutum*, *Spondias cyatherea*, *S. dulcis*, *S. mombin*, *S. purpurea*, *S. venosa*, *Tapirira guaianensis*, *Thyrosodium paraensis*, *Toxicodendron radicans*.

ANNONACEAE
128 genera; 2,050 species

This is a family of the Old World tropics, but *Asimina* is also found in temperate regions including the United States. The Annonaceae are familiar as a source of edible fruits throughout the world (custard apple, cherimoya, soursop, etc.).

At least 50 genera including some 75 species are known to be alkaloidal; benzylisoquinolines, aporphines, berberines, and a variety of other N-containing compounds are found throughout the family. Recent reviews of some of these constituents are available (Cave et al., 1989; Waterman, 1985; Zhong and Xie, 1988). The following record of positive alkaloid tests was obtained from 240 samples comprising 155 species.

These plants, known to be alkaloidal, were recognized: *Annona montana*, *A. muricata* (2/4), *A. reticulata*, *A. squamosa*, *Asimina triloba* (5/6), *Guatteria psilopus* (4/4), *Hexalobus monopetalus* (1/2), *Monodora myristica*, *Popowia pisocarpa*, *Rauwenhoffia leichthardtii*, *Rollinia mucosa* (8/9), *Trivalvaria pumila* (2/2), *Xylopia aethiopica*, *X. papuana*.

Other positive tests included *Alphonsea sp.*, *Annona arenaria* (3/3), *A. chrysophylla*, *A. crassiflora* (1/2), *A. exsucca*, *A. palustris*, *A. senegalensis* (5/7), *Annona spp.* (2), *A. stenophylla*, *Artabotrys monteirosae* (1/2), *A. odoratissima* (stem), *Asimina longifolia*, *A. nashii*, *A. parviflora*, *A. speciosa*, *Bocagegopsis multiflora*, *Cananga blainii* (4/4), *C. odorata* (5/6), *Cleistochlamys kirkii* (1/3), *Cleistopholis patens* (1/3), *Crematosperma polyphlebum*, *Cymbopetalum penduliflorum* (2/2), *Desmos sp.*, *Duguetia* aff. *amazonica*, *D. odorata*, *Duguetia spp.* (3), *D. spixina*, *D. surinamensis* (1/2), *Enneastemon schweinfurthii*, *Ephedranthus amazonicus*, *Ephedranthus spp.* (2), *Fusaea longifolia* (3/3), *Goniothalamus sp.*, *Guatteria blainii* (3/3), *G. calva*, *G. duckeana*, *G. dura* (bark), *G. elongata*, *G. megaphylla*, *G. micans*, *G. odorata* (bark, leaf), *Guatteria spp.* (7/18), *Monodora grandiflora* (2/3), *Oxymitra sp.*, *Papualthia spp.* (2), *Phaeanthus macropodus* (2/2), *Polyalthia armittana* (2/3), *P. glauca*, *P. oblongifolia* (2/2), *Polyalthia sp.*, *Popowia fusca*, *P. obovatum*, *Popowia sp.*, *Pseuduvaria spp.* (2/2), *Rollinia sp.*, *Tetrameranthus duckei*, *Unonopsis sp.*, *Uvaria chamae*, *Uvaria sp.*, *Xylopia amazonica* (4/4), *X. aromatica* (2/4), *X. ochrantha*, *X. sericea*, *Xylopia sp.*, *X. tomentosa*.

Negative tests were obtained for the following: *Anaxagorea dolichocarpa*, *Annona ambotay*, *A. dioica*, *A. globiflora*, *A. jiquitahi*, *A. longiflora*, *A. longipeps*, *A. menticola*, *Annona spp.* (11), *Artabotrys brachypetalus*, *Cyathocalyx ramuliflorus*, *C. ridleyi*, *Cymbopetalum brasiliense*, *Desmos dasymaschalus*, *Duguetia furfuracea*, *Guatteria australis*, *G. insculpta*, *G. meliodora*, *Guatteria spp.* (10), *Isolona campanulata*, *Miliusa velutina*, *Mitrella* (= *Fissistigma*) *kentii*, *Oxandra lanceolata*, *Polyathia sp.*, *Rollinia dolabripetala*, *R. exalbida*, *R. exsucca*, *R. laurifolia*, *Rollinia spp.* (6), *Saccopetalum* (= *Miliusa*) *tomentosum*, *Uvaria afzelii*, *Xylopia benthami*, *X. frutescens*, *X. grandiflora*, *X. lingustifolia*, *X. longsdorfiana*, *X. malayana*, *Xylopia spp.* (7).

REFERENCES

Cave, A., M. Lebeouf, and B. K. Cassels, *Alkaloids* (Academic Press) *35* (1989) pp. 1-76.

Cave, A., M. Lebeouf, and P. G. Waterman, *Alkaloids: Chemical and Biological Perspectives 3* (1985) pp. 133-270.

Zhong, S. and Xie, N., *Zhongguo Yaoke Daxue Xuebo 19* (1988) p. 156 (Chem. Abs. 109: 98613u).

APOCYNACEAE
215 genera; 2,100 species

This family is almost cosmopolitan but chiefly tropical with some representatives in the temperate zones. It is noted for many ornamentals, some species that yield rubber (*Funtumia*, *Landolphia*), and several useful drugs (*Strophanthus*, *Acokanthera*, *Catharanthus*, *Rauvolfia*). Many are toxic.

The Apocynaceae is probably the most thoroughly investigated family for alkaloidal plants; about 1,000 of these compounds have been isolated from its many members. This intense interest followed the isolation and characterization of reserpine and its relatives from the traditional Indian drug *Rauvolfia serpentina* and the discovery of the antileukemic alkaloids of *Catharanthus*. The study reported here was done, in part, during the time these events took place and, as a result, a degree of emphasis was placed on screening "unusual" representatives of the family. Some 775 samples including 443 species were examined.

Many of the species recognized as alkaloidal by other investigators were confirmed as such: *Allamanda cathartica* (4/7), *Alstonia macrophylla*, *Alyxia olivaeformis*, *A. ruscifolia*, *Aspidosperma discolor*, *A. macrocarpon*, *A. megalocarpon*, *A. pyrifolium*, *Catharanthus roseus* (4/4), *Diplorhynchus condylocarpon* (6/7), *Ervatamia dichotoma* (5/5), *Funtumia africana* (1/2), *F. elastica* (8/8), *F. latifolia* (3/5), *Gabunia odoratissima* (4/9), *Geissospermum vellozii*, *Haplophyton cimicidum*, *Holarrhena febrifuga*, *H. wulfsbergii*, *Macoubea guianesis* (2/2), *Malouetia arborea*, *M. tamaquarina* (6/7), *Nerium oleander*, *Ochrosia elliptica* (2/4), *Pagiantha cerifera* (2/2), *Parsonsia straminea*, *P. velutina* (1/2), *Peschierea affinis* (6/6), *Pleiocarpa mutica* (8/8), *Rauvolfia caffra*, *R. chinensis*, *R. degneri*, *R. hirsuta*, *R. mauiensis*, *R. tetraphylla*, *R. verticillata*, *R. viridis*, *R. vomitoria* (2/2), *Rhazya stricta* (5/5), *Stemmadenia donnellsmithii* (2/2), *S. galeottiana* (1/2), *S. obovata* (3/4), *Tabernaemontana dichotoma*, *T. elegans*, *T. pandacqui*, *T. ridelii*, *T. rigida*, *T. rupicula* (3/3), *Tonduzia longifolia* (2/2), *Vallesia glabra* (4/4), *Vinca major*, *V. minor*, *Voacanga thoursii*, *Wrightia tomentosa* (fruit).

In addition, the following were positive: *Allamanda violacea* (1/2), *Alstonia boonei* (2/3), *A. congoensis*, *A. costata*, *A. glabriflo-*

ra, A. montana, Alstonia spp. (2/4), *A. vitensis* (2/2), *Alyxia annamensis, A. buxifolia, A. concatenata* (1/2), *Alyxia* cf. *markgrafia* (2/2), *A. flavescens, A. fragrans* (1/2), *A. lanceolata, A. laurina, A. loesseriana, A. lucida* (1/2), *A. punctata, A. sinensis* (1/2), *A. spicata, A. zeylanica, Ambelania sp., Amsonia breviflora, Apocynum androsaemifolium, A. camporum, Aspidosperum cruentum* (bark, fruit), *A. olivaceum, Aspidosperma spp.* (4/5), *Baissea wulfhorstii, Beaumontia grandiflora* (3/5), *Beaumontia sp., Bonafusia hirtula, B. sananho, B. tetrastachya, B. undulata, Carissa bispinosa* (1/2), *C. grandiflora* (1/2), *C. lanceolata* (1/2), *Carruthersia carruthersia, Clitandropsis* (= *Melodinus*) *novoguinensis, Conopharyngia elegans, C. holstii* (2/2), *Ervatamia eriophora* (1/2), *Hunteria africana* (3/3), *H. zeylanica* (2/2), *Kopsia fruticosa* (2/2), *Landolphia kirkii, Lepiniopsis ternatensis* (2/2), *Macoubea guianesis* (2/2), *Malouetia furfuracea* (3/3), *M. nitida, Mandevilla cuneifolia, M. illustris* (whole plant), *Marsdenia rubrofusca, Melodinus landolphoides, M. monogynus* (2/2), *Ochrosia sandwicensis* (2/2), *Odontodenia sp., Pagiantha dichotoma* (1/2), *P. heyneana, P. oligantha, P. plumeriafolia, P. sphaerocarpa, P. subglobosa, P. thurstonii, Parsonsia albiflora, P. helicandra, Peschierea australis, P. bahia, P. laeta, Peschierea spp.* (6/7), *Plumeria rubra, Prestonia mexicana* (2/2), *Prestonia sp., Rauvolfia heterophylla* (2/2), *R. oxyphylla, R. sandwicensis, R. semperflorens* (whole plant) *R. suaveolens, Stemmadenia ebracteata, S. grandiflora, S. palmeri, Stemmadenia spp.* (4/4), *Strophanthus gratus, Tabernaemontana alba* (1/3), *T. angulata, T. barteri* (2/2), *T. citrifolia* (3/3), *T. crassa, T. grandiflora, T. littoralis* (2/2), *T. maxima, T. muricata* (2/2), *T. oblongifolia, T. pacifica, T. psychotrifolia, T. sananho* (3/3), *Tabernaemontana spp.* (6/9), *T. stenoloba, T. submollis* (2/2), *Thevetia peruviana, Trachelospermum fragrans* (bark), *Trachelospermum jasminoides, Urceola brachycephala, Vinca lancea* (2/2), *V. pusilla, Vinca spp.* (2/2), *Voacanga natalensis, Wrightia pubescens.*

Negative tests were obtained from the following: *Acokanthera oblongifolia, A. oppositifolia, A. schimperi, Adenium multiflorum, Aganosma acuminata, A. aganosma, A. caryophyllata, A. cymosa, A. gracile, A. marginata, A. schlechteriana, A. velutina, Allamanda spp.* (3), *Alstonia macrophylla, Alstonia spp.* (2), *Alyxia acutifolia, A. affinis, A. amoena, A. arfakensis, A. bodinieri, A. bracteolosa, A. brevipes, A. cacumina, A. celastrina, Alyxia* cf. *defoliata, Alyxia* cf. *pullei, A. clusiophylla, A. disphaerocarpa, A. doratophylla, A. elliptica, A. erythrosperma, A. floribunda, A. forbesii, A. glaucophylla, A. hainanensis, A. ilicifolia, A. intermedia, A. lamii, A. lata, A. laxiflora, A. levinei, A. leucogyne, A. linearifolia, A. luzonensis, A. microbuxus, A. monticola, A. myrtillaefolia, A. mummularia, A. orophila, A. parvifolia, A. pisiformis, A. pseudosinensis, A. purpureoclada, A. reinwardii, A. revoluta, A. romarinifolia, A. schlechteri, A. scabrida, A. scandens, A. serpentina, A. sibuyanesis, A. sorgerensis, Alyxia spp.* (2), *A. stellata, A. subalpina, A. torqueata, A. torresiana, A. yunkuniana, Anchornia sp., Ancylobothrys petersiana, Angadenia berteri, A. lindeniana, Anodendron affine, A. axillare, A. benthamianum, A. candolleanum, A. coriaceum, A. laeve, A. loheri, A. manubriatum, A. oblongifolium, A. paniculatum, A. punctatum, Apocynum cannabinum, A. sibricum, Artia orbicularis, Aspidosperma polyneuron, Aspidosperma sp., Cameraria angustifolia, C. belizensis, C. latifolia, C. longii, C. retusa, Carissa edulis, C. haematocarpa, C. macrocarpa, Carruthersia brassii, C. daronensis, C. macgregorii, C. pilosa, C. latifolia, C. scandens, Cerbera floribunda, C. odallam, Cerbera spp.* (3), *Clitandra orientalis, Condylocarpon rauvolfiae, Couma macrocarpa, C. utilis, Ecdysanthera rosea, Echites umbellata, Forsteronia leptocarpa, F. luschnathii, F. riedelii, F. rufa, Forsteronia spp.* (3), *F. thyrsoidea, Himatanthus articulatus, H. bracteatus, H. obovataus, Himatanthus spp.* (14), *H. subcarnosa, H. steyermarkii, Ichnocarpus frutescens, Kopsia flavida, Landolphia buchanalii, L. capensis, L. owarensis, L. ugandensis, Laseguea erecta, Lyonsia reticulata, Macrosiphonia brachysiphon, M. hypoleuca, M. longiflora, M. macrosiphon, M. martii, M. petraea, Macrosiphonia spp.* (5), *M. velame, Mandevilla filiformis, M. foliosa, M. funiformis, M. immaculata, M. karwinskii, M. lesigna, Mandevilla spp.* (17), *M. steyermarkii, M. subcarnosa, M. subsaggitata, Melodinus baueri, M. suaveolens, Mesechites trifida, M. trifolia, Nerium indicum, N. odorum, Odontodenia grandiflora, Pachypodium leolii, P. saundersii, P. succulentum, Pagiantha macrocarpa, P. megacarpa, P. pandacqui, Parahancornia sp., P. peruviana, Parsonsia baudoinii, P. brassii, P. brunensis, P. canescens, P. capsularis, P. carnea, P. contusa, P. crebriflora, P. cummingiana, P. curvisepala, P. edulis, P. fulva, P. heterophylla, P. javanica, P. laevis, P. lata, P. lilacina, P. molissi-*

ma, P. rotata, P. rubra, Parsonsia sp., P. ventricosa, Peltastes sp.,
Peschierea australis, Plumeria acutifolia, P. obtusa, Plumeriopsis
ahouai, Pottsia grandiflora, P. laxiflora, P. ovata, Prestonia acuti-
folia, P. agglutinata, P. amanuensis, P. bahiensis, P. brachypoda,
P. coalita, P. concolor, P. guatamalensis, P. hassleri, P. isthmica,
P. lindleyana, P. lindmannii, P. marginata, P. mollis, P. obovata,
P. peregrina, P. portobellensis, P. quinquangularis, P. riedelii,
P. solanifolia, Prestonia spp. (5), *P. tomentosa, P. trifida, Pteralyxia*
macrocarpa, Rauvolfia linearisepala, R. sellowii, Rhabdadenia bi-
color, R. biflora, R. macrostoma, Saba florida, Secondatia densiflo-
ra, Skytanthus sp., Stipecoma peltata, Strophanthus gerardii,
S. hispidus, S. luteolus, S. petersianus, S. sarmentosus, S. speciosus,
S. welwitchii, Tabernaemontana heyneana, Tabernaemontana spp.
(4), *Temnadenia sp., T. stellaris, Thevetia neriifolia, T. ovata,*
T. peruviana, T. thevetoides, Urceola javanica, U. lucida, U. philip-
pinensis, U. torulosa, Urechites andrieuxii, U. lutea, Vallaris hey-
nei, V. solanacea, Wrightia saligna, Wrightia sp.

Several samples of the less common genera in this extensive
family were supplied in the form of gleanings from herbarium
specimens.

AQUIFOLIACEAE
4 genera; 420 species

Most of the species in this family are in the genus *Ilex*, which has
three centers of distribution: South America, North America, and
the South Pacific. The genus is important as a source of lumber and
ornamentals (holly) and, in South America, as a basis for traditional
caffeine-containing drinks (mate and guayusa).

The chemistry of the family is that of the major genus (*Ilex*)
known for its content of caffeine and theobromine along with cyano-
glucosides of a sort which do not liberate HCN on usual hydrolysis.

Of the 42 species of *Ilex* tested, only four were regarded as
positive: *I. cassine, I. coriacea, I. crenata,* and *I. glabra* (2/3).
I. cassine and *I. crenata* had previously been reported as alkaloid-
positive. The purines do not give definitive alkaloid tests with the
Dragendorff reagent and are not considered true alkaloids by some
investigators.

The following species were negative: *Ilex anomala, I. arnhemica, I. asprella, I. bioritensis, I. burfordii, I. chamaedryfolia, I. cornuta, I. discolor, I. diuretica, I. dumosa, I. formosana, I. hanceana, I. impressivena, I. incana, I. incarnata, I. jennanii, I. laevigata, I. microdonta, I. mitis, I. opoca, I. paraguariensis, I. parvifolia, I. pubescens, I. rotunda, I. serrata, Ilex spp.* (5), *Ilex* cf. *versteeghii, I. verticillata, I. verticillata, I. vitis-idaea, I. vomitoria.*

Phelline comosa, Sphenostemon arfakensis, Sphenostemon cf. *arfakensis,* and *S. papuanum* were also negative; they are sometimes placed in families of their own, Phellinaceae and Sphenostemonaceae respectively.

ARACEAE
106 genera; 2,950 species

The family is mostly tropical and subtropical but extends into temperate areas including a few representatives in the United States. In the New World, we recognize some genera as ornamentals (e.g., the calla lily); in the Old World, some roots are used as food (e.g., taro) as are the fruits of *Monstera* species.

Alkaloids are known for some 25 genera (35 species) in the family; coniine, hydroxytryptamine, berberines, and an assortment of other N-containing compounds has been identified.

Ninety-four samples representing 73 species were examined in this study. *Colocasia esculenta, Symplocarpus foetidus,* and *Zantedeschia aethiopica* (1/5) had been previously reported as alkaloid-positive.

Several other species were found here to give positive tests as well: *Acorus calamus, A. gramineus, Alocasia odora, Anthurium sp.* (1/3), *Cyrtosperma johnstoni, Peltandra virginica, Symplocarpus foetidus, Typhonium divaricatum, Urospatha saggitaefolium* (1/2), *Zantedeschia melanoleuca, Z. rehmannii.*

Some of the literature reports of the presence of alkaloids in this family may have resulted from the use of ammonium hydroxide during isolation. This practice has been shown to convert certain of the plant constituents to N-containing compounds, which then react as alkaloids in standard testing procedures.

Negative tests were obtained with the following species: *Acorus gramineus* var. *pusillus, Aglaonema modestum, Alocasia indica, Amorphophallus glabra, A. montrichardia, Anthurium mexicanum, A. pedatoradiatum, Anthurium spp.* (2), *A. scandens, Arisaema draconitum, A. japonicum, A. triphyllum, Arum maculatum, Calla sp., Colocasia antiquorum, Epipremum pinnatum, Heteropsis sp., Lasia spinosa, Monstera pertusa, Montrichardia* (= *Amorphophallus*) *arborescens, Montrichardia sp., Orontium aquaticum, Philodendron imbe, P. inaequilaterum, P. obliquifolium, P. rudgeanum, P. sequine, P. selloum, Philodendron spp.* (7), *Pistia stratioides, Pothos seemanni, Rhodospatha roseospadix, Richardia* (= *Zantedeschia*) *brasiliensis, R. scabra, Spathiphyllum cochlearispathum, Spathiphyllum sp., Stylochiton natalensis, S. puberulus, Stylochiton sp., Synantherias* (= *Amorphophallus*) *sylvatica, Syngonium llamasii, S. podophyllum, Syngonium sp., S. vellosianum, Urospatha sp., Xanthosoma mendozae, X. mexicanum, X. robustus, X. saggitifolium, X. violaceum.*

ARALIACEAE
57 genera; 800 species

This is primarily a tropical family with centers of distribution in Indo-Malaysia and tropical America. Three genera are found in the United States.

English ivy and others are cultivated as ornamentals; some are used as medicine (the traditional Chinese drug ginseng belongs in this family).

Several unnamed alkaloids have been recorded in some ten genera of the family. Quinazolines have been characterized from the genus *Mackinlaya*; the known alkaloidal *M. schlechteri* was also found positive in this study.

Positive tests were likewise obtained with the following species: *Aralia racemosa* (1/4), *Cussonia paniculata, C. thyrsiflora, C. umbellifera* (1/3), *Dendropanax pellucipunctata* (1/5), *Didymopanax tremulum* (1/2), *Gastonia papuana, Seemannaralia gerrardii.*

The following species were negative: *Acanthopanax trifoliatus, Aralia californica, A. hispida, A. humilis, A. nudicaulis, A. regeliana, A. spinosa, Astrotricha asperifolia, A. flocosca, Brassaia actinophylla, Cussonia kirkii, C. natalensis, C. spicata, Dendropanax arboreum, D. cuneatum, D. parviflorum, D. pellucidopunctatus, Didymopanax angustissium, D. morototoni, Didymopanax spp.* (3), *D. vinosum, Dizygotheca* (= *Schefflera*) *coenosa, Dizygotheca sp., Gilibertia arborea, G. cuneata, Hedera helix, Heptapleurum* (= *Schefflera*) *arboricolum, H. octophyllum, H. venulosum, Kalopanax* (= *Eleutherococcus*) *pictus, Kissodendron* (= *Polyscias*) *australianum, Mackinlaya macrosciadia, Meryta sp., Myodocarpus sp., Neopanax arboreum, N. colensoi, N. simplex, Oreopanax capitatum, O. echinops, O. fulvum, O. salvinia, Oreopanax spp.* (2), *O. xalapensis, Panax ginseng, Plerandra* (= *Schefflera*) *stahliana, P. vitiensis, Polyscias balfouriana, P. elegans, P. filicifolia, P. guilfoylei, P. sambucifolia, Pseudopanax crassifolium, P. edgerleyi, P. lessonii, Schlefflera digitata, S. octophylla, S. taiwaniana, Tetrapanax papyriferus, Tetraplasandra* (= *Gastonia*) *sp., Tieghemopanax* (= *Polyscias*) *elegans.*

ARAUCARIACEAE
2 genera; 32 species

Members of this family are ornamental southern pines familiar to horticulturists. In the southern hemisphere, except in Africa and southeast Asia, some are the source of lumber and resins.

There has been but one positive test for alkaloids recorded for this small gymnosperm family (*Agathis australis*). A test of this species was negative as well as tests on *A. moorei, A. ovata, A. robusta, Agathis sp., A. vitensis, Araucaria bididellii, A. cookii, A. cunninghamii, A. excelsa,* and *A. rulei.*

ARISTOLOCHIACEAE
7 genera; 410 species

This is essentially a tropical family but some representatives occur in the temperate zone.

Nitrophenanthrenes and their reduced (amino) counterparts as well as quarternary aporphines are characteristic. Some species have been used as medicinals.

Positive tests for alkaloids were obtained with the following species previously known to be alkaloidal: *Aristolochia elegans*, *A. gigantea*, *A. tagala*. In addition, an unidentified *Aristolochia sp.* was found to be positive (1/3).

Negative results were obtained for *Aristolochia burchellii*, *A. didyma*, *A. jaliscana*, *A. kankauensis*, *A. macrophylla*, *A. paulistana*, *Aristolochia spp.* (3), *A. triangularis*, *Asarum canadense*, *A. lemonii*, *A. taitoense*.

ASCLEPIADACEAE
347 genera; 2,850 species

Although this family is pantropical, most of its members are South American. A few genera extend into temperate regions; one of these is the familiar milkweed, *Asclepias syriaca*. Some are ornamentals, some yield rubber, others are livestock poisons. The taxonomy of the family is not a matter of general agreement.

Few alkaloids have been found in this relatively large family. Those in *Cryptolepis*, *Cynanchum*, *Pergularia*, *Tylophora*, and *Vincetoxicum* have been characterized, some have been synthesized, others have yet to be isolated in pure form.

In this study, 182 species were tested with the following previously known alkaloidal plants found positive: *Asclepias curassavica* (2/13), *A. linaria* (2/5), *Calotropis gigantea* (1/3), *Ectadiopsis oblongifolia* (1/2), *Marsdenia condurango*.

These species were also positive: *Asclepias cordifolia* (1/2), *A. fasciculata*, *A. fruticosa*, *A. humistrata*, *A. rotundifolia*, *A. subverticillata* (2/4), *A. vestita*, *Blepharodon sp.*, *Caralluma mammiliaris*, *Caralluma sp.*, *Cryptolepis oblongifolia* (1/2), *Cynanchum mitreola*, *C. nigrum*, *C. praecox*, *Gomphocarpus physocarpus*, *Gonolobus gonocarpus*, *G. obliquus*, *Hemidesmus indicus* (1/2), *Heterostemma collinum*, *H. papuana*, *Hoodia sp.*, *Kanahia laniflora* (1/2), *Margaretta rosea*, *Marsdenia dregei*, *M. rostrata*, *Microloma incanum*, *M. massonii*, *M. saggitatum* (1/2), *Pachycarpus rigida*, *P. scaber*, *Pectinaria breviloba*, *Pentarrhinum insipidum*, *Pergularia daemia-extensa* (3/10), *Pergularia sp.*, *Secamone garardii*, *Stapelia gigantea* (2/3), *S. olivacea*, *S. schinzii*, *Stapelia sp.*, *Toxocar-*

pus wrightianus, Tylophora macrophylla, T. ovata, Xysmalobium undulatum.

The following species gave negative tests: *Araujia sericofera, Asclepias albicans, A. amplexicaulis, A. angustifolia, A. auriculata, A. bidentata, A. brachystephania, A. burchellii, A. californica, A. contrayerba, A. filiformis, A. gibba, A. glaberrima, A. glaucescens, A. incarnata, A. melantha, A. mexicana, A. neglecta, A. oenotheroides, A. orata, A. oratoides, A. ovata, A. pringlei, Asclepias spp.* (2), *A. speciosa, A. subulata, A. syriaca, A. verticillata, Aspidoglossum biflorum, Blepharodon mucronatum, B. steudelianum, Brachystelma pygmaeum, Caralluma picranthoides, C. Ceropegia abyssinica, C. occulta, Chlorocodon* (= *Mondia*) *whitei, Cosmostigma racemosum, Cryptolepis capensis, C. cryptolepoides, Cynanchum africanum, C. ellipticum, C. floribundum, C. freemani, C. kunthii, C. obtusifolium, C. parviflorum, C. pringlei, Dischidia rafflesiana, Dischidia sp., Ditassa acerosa, D. ridelii, D. edmundoi, Ditassa sp., Dregea abyssinica, Finlaysonia obovata, Fockea lugardii, F. multiflora, Glossostelma carsonii, Gomphocarpus* (= *Asclepias*) *aureus, G. glaucophyllus, Gomphocarpus sp., Gonolobus chrysanthus, G. broadwayi, G. pilosus, G. productus, G. uniflorus, Gymnema laterniflorus, G. sylvestre, Hoya bicarinata, Marsdenia hilariana, M. macrophylla, M. mexicana, M. pringlei, M. rubrofusca, Matelea hirsuta, M. pavonii, Matalea sp., Metastelma* (= *Cynanchum*) *angustifolium, Metastelma sp., Microloma tennuifolium, Mondia whitei, Oleandra wallichii, Orthosia urceolata, Oxypetalum arnottianum, O. banksii, O. panosum, O. pedicillatum, Oxypetalum spp.* (9), *O. sublanatum, Pachycarpus appendiculatus, P. validus, Pentatropis cynanchoides, Pergularia spp.* (2), *Pilostigma* (= *Constantina*) *thonningii, Raphionacme burkei, R. elata, R. flanagani, R. hirsuta, Riocreuxia picta, R. torulosa, Sarcolobus clausum, Sarcolobus* cf. *globosus, S. elegans, S. viminale, S. mosensii, Sarcostemma spp.* (3), *Schistogyne sp., Schizoglossum petherickanum, Secamone albinii, S. frutescens, S. parvifolia, S. gettleffii, Stapelia variegata, Stomatostemma monteiroae, Stultitia* (= *Orbea*) *tapscottii, Tacazzea apiculata, Tassadia propiniqua, Tylophora grandiflora, Vincetoxicum sp.*

The following genera have been placed in a separate family, Periplocaceae, by some taxonomists: *Finlaysonia, Hemisdesmus, Mondia, Raphionacme, Stomatostemma, Tacazzea.*

B

BALANITACEAE
1 genus; 25 species

The genus *Balanites* of tropical Asia and Africa is now listed among the Zygophyllaceae by some authorities. The seeds of some species yield oils used in soap making; others are medicinal.

One sample of an undetermined species gave a positive test for alkaloids in this study. The chemistry of the family is otherwise unknown, although alkaloids are known for the Zygophyllaceae.

BALANOPACACEAE
1 genus; 9 species

This small family is native to the southwest Pacific including Queensland in Australia. It has no economic uses.

One species, *Balanops australiana*, gave a negative test for alkaloids. Nothing is known of the chemistry of the family.

BALSAMINACEAE
2 genera; 850 species

These plants are widely distributed but are most abundant in the Asian and African tropics. Several species are ornamentals.

Alkaloids have been reported for two species of *Impatiens*, but 12 samples including the following 11 species were tested in this survey without positive results: *Impatiens biflora*, *I. cecili*, *I. chinensis*, *I. duthiei*, *I. kirkii*, *I. pallida*, *Impatiens spp.* (3), *I. sylvicola*, *I. uniflora*.

BASELLACEAE
4 genera; 15 species

This family is found mostly in tropical America and the West Indies, with one species native to Asia. Some are cultivated as ornamentals; others are used for foods–a leafy vegetable (*Basella*) and a starchy root of the Andes (*Ullucus*).

Eight samples representing five species gave negative tests for alkaloids, which have not yet been found in the family: *Anredera vesicaria, Basella rubra, Boussingaultia* (= *Anredera*) *baselloides, B. leptostachys, B. ramosa.*

BATACEAE (BATIDACEAE)
1 genus; 2 species

This is a family of the shorelines of the tropics and subtropics of the New World and Hawaii. It is of no known economic importance.

Indolic glucosinolates have been reported in *Batis maritima*, but a test of this species did not give a reaction with Dragendorff's reagent.

BEGONIACEAE
2 genera; ca. 900 species

The family has wide distribution throughout the tropics, especially in South America. Varieties of many species have been developed by horticulturists and grown as familiar garden and house plants.

Alkaloids are not known in the family. Twenty-six samples representing 21 species were tested without positive result: *Begonia balsaminea, B. caffra, B. fruticosa, B. gracilis, B. heracleifolia, B. hispida, B. incarnata, B. inciso-serrata, B. macdougallii, B. nelumbiifolia, B. palmaris, B. princeae, B. randaiensis, B. ricinifolia, B. scandens, Begonia spp.* (4), *B. tovarensis, B. ulmifolia.*

BERBERIDACEAE
15 genera; 570 species

A few members of the family are found in South America but as a group the Berberidaceae are chiefly north temperate. Many are used as ornamentals and some bear edible fruit.

Earlier taxonomists included 12 genera in the family, 11 of which have been reported to contain alkaloids in 86 species. Thirty-two samples of 16 species were tested in this study. The following were found to be alkaloidal as reported in the literature: *Berberis laurina, B. thunbergii, B. trifoliata, B. vulgaris* (2/2), *B. wallichiana, Caulophyllum thalictroides, Mahonia citrifolia, Nandina domestica* (2/3).

These species were also alkaloid-positive: *Berberis incertus, Epimedium violaceum, Podophyllum peltatum* (3/3).

Caulophyllum, Nandina, and *Podophyllum* are each given separate family status by some authorities.

Achlys triphylla, Berberis gracilis, B. lanceolata, and a *Berberis sp.,* along with a sample of *Mahonia ilicina,* were negative.

BETULACEAE
6 genera; 150 species

Most of the members of this family are found in the northern hemisphere. Several are known for excellent hardwood (e.g., birch) as well as edible nuts (e.g., hazelnuts, filberts). *Carpinus* is sometimes placed in a family of its own as is *Corylus* (see Corylaceae).

Positive alkaloid tests have been reported for two species of *Alnus* and one of *Betula.* Of 44 samples tested here the following positive results were obtained: *Alnus glabrata* (1/5), *A. rhombifolia, A. rubra, Betula papyifera, Carpinus japonica.*

Most of the species tested were negative: *Alnus arguta, A. firmifolia, A. formosana, A. jorullensis, A. pringlei, A. rugosa, A. serrulata, Alnus sp.* (2), *Betula aurata, B. lenta, B. lutea, B. mandshuria szechuanica, B. nana, B. nigra, B. populifolia, Carpinus caroliniana, C. kawakami, C. minutiserrata, C. rankanensis, C. tropicalis.*

BIGNONIACEAE
112 genera; 725 species

Primarily a tropical family with representation especially in northern South America, the Bignoniaceae furnish lumber and many ornamentals.

There are records of alkaloids in 24 genera and as many species.

In this study, 212 samples in 126 species were tested; three species had been known to be alkaloidal: *Campsis radicans* (1/3), *Catalpa bignonioides* (1/4), and *Stenolobium* (= *Tecoma*) *stans*. In addition, the following were also positive: *Astianthus viminalis* (1/2), *Bignonia anomotegma, Callichlamys latifolia* (2/2), *Catalpa ovata, Cybistax antisyphilitica, Cybistax sp.,* C. *donnell-smithii, Cydista aequinoctialis* (1/2), *Deplanchea spinosa, Deplanchea sp., Dolichandrone alba, D. spathacea* (1/3), *D. falcata, Enallagma* (= *Dendrosicus*) *latifolia, Kigelia moosa, K. pinnata* (1/4), *Markhamia platycalyx, Martinella obovata, Pandorea doratoxylon, Podranea brycei* (2/2), *Rhigozum brevispinosum* (1/2), *Spathodea campanulata* (1/6), *S. nilotica, Tabebuia fluviatilis, T. pentaphylla, T. serratifolia* (1/2), *Tabebuia spp.* (2), *Tecoma sp.* (1/2), *T. tronadora, Tecomanthe sp., Tecomaria capensis* (2/3), *Zeyheria tuberculosa* (1/2).

Negative results were obtained with the following: *Adenocalyma comosa, A. heterophylla, Adenocalymna spp.* (3), *Amphilophium molle, A. paniculata, Anemopaegma prostratum, Anemopaegma sp., Arrabidea brachypoda, A. chica, A. claussenii, A. japurensis, A. littoralis, A. nigrescens, A. mollis, Arrabidea spp.* (2), *Bignonia angus-castus, B. anomotegma, B. capreolata, B. exoleta, Bignonia sp., B. unguiculata, B. venusta, Catophractes alexandri, Chilopsi linearis, Clytostoma sciuripablum, Crescentia alata, C. amazonica, C. cujete, C. latifolia, Cuspidaria pterocarpa, Deplanchea tetraphylla, Doxantha* (= *Macfadyena*) *unguiculata, Fridericia sp., F. speciosa, Haplophragma* (= *Fernandoa*) *adenophyllum, Heterophragma adenophyllum, Jacaranda copaia, J. micrantha, J. mimosaefolia, J. obtusifolia, J. oxyphylla, J. puberula, J. semiserrata, Jacaranda spp.* (4), *Kigelia aethiopica, Leucocalantha aromatica, Lundia corymbifolia, L. nitidula, Markhamia acuminata, M. hildebrandtii, M. obtusifolia, Memora flavida, M. glaberrima, Newbouldia sp., Oroxylon indicum, Pandorea pandorana, Paragonia pyramidata, Parmentiera cerifera, P. edulis, Petastoma* (= *Arrabidea*) *samynoides, Phryganocydia corymbosa, Pithecoctenium echinatum, Pithecoctenium sp., Pseudocalyma* (= *Mansoa*) *aliaceum, Pyrostegia ignea, P. venusta, Rhigozum obovatum, R. trichotomum, Schlegelia ramizii* (now in Scrophulariaceae), *Sparattosperma vernicosum, Spathodea xylocarpa, Stereospermum suaveolens, Sti-*

zophyllum perforatum, Tabebuia alba, T. barbata, T. caraiba, T. cassinoides, T. palmeri, T. rosea, T. stenocalyx, Tecoma austro-caledonica, T. chrysantha, Tecoma sp., T. undulata, Tynanthus micranthus, Zeyheria montana.

BIXACEAE
3 genera; 16 species

Bixa orellana of the American tropics is known for the red dye made from the testa around the seed, which is used as a food coloring and as a body paint among South American Indian peoples.

Alkaloids are not known, nor were they detected, in testing of six samples of this species.

BOMBACACEAE
30 genera; 250 species

Found chiefly in the American tropics, this family is the source of kapok, balsa wood, some ornamental trees, and the popular Asian fruit, durian.

Only three genera and four species had been reported to give positive alkaloid tests until the appearance of a series of papers on the chemistry of *Quararibea funebris* (Zennie and Cassaday, 1990) and a brief survey of the folk-medical uses of several genera (Raffauf and Simon, 1988). The results of this last study, which included extensive testing of herbarium specimens, are repeated here for a more complete survey of the family. A total of 349 samples representing 135 species were tested.

The following gave positive Dragendorff tests: *Bombax brevicuspe, B. buonopotenze* (1/2), *B. ellipticum* (1/5), *Ceiba aesculifolia* (1/5), *Cullenia rosayroana* (1/2), *Durio acutifolia* (2/3), *D. singaporensis* (2/3), *Ochroma bicolor* (2/2), *O. boliviana* (1/2), *O. lagopus* (3/7), *O. limonensis* (1/3), *O. pyramidale* (3/10), *Pachira aquatica* (1/2), *P. orinocense* (stem), *Quararibea funebris, Q. guianensis* (1/2), *Q. lasiocalyx, Q. ochrocalyx* (1/2), *Rhodognaphalopsis nitida, Spirotheca passifloroides.*

Negative tests were obtained with the following: *Adansonia digitata, Bernoullia flammea, B. seveitenoides, Bombax barrigon, B. blancoanum, B. ceiba, B. coriaceum, B. cumarrerrae, B. cumanense, B. cyathophorum, B. discolor, B. faroense, B. flammeum, B. flaviflorum, B. glabra, B. globosum, B. gracilipes, B. humile, B. insigne, B. jenmanni, B. longipedicillatum, B. macrophyllum, B. malabaricum, B. margitanum, B. mexicanum, B. millei, B. minus, B. munguba, B. nervosum, B. obtusum, B. palermi, B. parviflorum, B. pentaphyllum, B. poissonianum, B. pubescens, B. quinatum, B. rigidifolium, B. sclerophyllum, B. septenatum, B. sessile, B. sordidum, B. surinamense, B. tomentosum, B. trinitensis, B. vurrenbaquense, Bombacopsis emarginatum, B. orinocensis, B. quinatum, B. sessilis, Camptostemon philippense, C. schultzii, Catostemma commune, C. fragrans, C. sclerophyllum, Ceiba acuminata, C. aesculifolia, C. caesaria, C. erianthos, C. grandiflora, C. mandonii, C. pallida, C. pavifolia, C. pentandra, C. pubiflora, C. rivieri, C. solmonea, C. samauma, C. schottii, C. tomentosa, Chorisia crispiflora, C. insignis, C. integrifolia, C. pubiflora, C. speciosa, Coelostegia borneensis, C. griffithii, C. ramealis, Cullenia ceilanica, Durio carinatus, D. conicus, D. dulcis, D. grandiflorus, D. graveolens, D. griffithii, D. insipida, D. kinabaluensis, D. kutejenias, D. lanceolatus, D. lowianus, D. macrophyllus, D. malaccensis, D. mansonii, D. natalensis, D. oblongus, D. oxleyanus, D. puteh, D. scortechinii, D. testudinarius* var. *pinangianus, D. zibethinus, Gossampinus malabarica, Matisia paraensis, Neesia altissima, N. glabra, N. malayana, N. pilulifera, N. syandra, Ochroma conolor, O. grandiflora, O. peruviana, O. tomentosa, O. velutina, Pachira emarginata, P. fatuosa, P. insignis, P. macrocarpa, Patinoa ichthyotoxica, Quararibea asterolepis, Q. bracteolosa, Q. petrocalyx, Q. putamayensis, Q. turbinata, Scleronema micranthum, Septotheca tessmannii.*

In an attempt to use the chemical constituents of the *Quararibea-Matisia* complex as an aid in their classification, these simple testing procedures were applied to 20 Central American samples representing either of the two genera. The highly odorous furanones, as well as positive alkaloid tests, were noted in 13 of these samples and absent in seven. The suggestion is that the former represent

Quararibea species, the latter *Matisia*. Classification based on the usual morphological criteria is under way as this is being written.

REFERENCES

Raffauf, R.F. and H.J. Simon, *Social Pharmacology 2* (1988) pp. 273-283.
Zennie, T.M. and J.M. Cassady, *Journal of Natural Products 53* (1990) 1611-1614 and references therein.

BORAGINACEAE
54 genera; 2,500 species

This is a family of wide distribution. It has economic use as a source of ornamentals; some species yield timber, and a dye is obtained from *Alkanna tinctoria*.

Pyrrolizidine alkaloids are among those identified in 73 species in 27 genera of the family. The 284 samples tested in this study included 176 species, two of which, *Heliotropium supinum* and *Trichodesma africanum*, were known to have been alkaloidal. Others giving positive tests included *Cordia cylindrostachya* (1/3), *C. monoica*, *C. scleriana* (1/2), *C. sebosteva*, *C. verbenacea* (1/2), *Cynoglossum australe*, *Ehretia tenuifolia* (considered by some taxonomists to be in a family of its own, Ehretiaceae), *Heliotropium angustifolium* (1/4), *H. curassavicum* (2/5), *H. gibbosum*, *Heliotropium supinum*, *Mertensia virginica* (1/2), *Onosma tauricum*, *Tournefortia* aff. *penniana*, *T. floribunda*, *Trichodesma africanum*, *T. zeylancium*.

Negative tests were obtained for the following: *Alkanna tinctoria*, *Amsinckia intermedia*, *A. menziesii*, *A. spectabilis*, *Anchusa capensis*, *Antiphytum heliotropoides*, *Borago officinalis*, *Bothriospermum tenellum*, *Bourreria obovata*, *Bourreria sp.*, *B. spathulata*, *Coldenia canescens*, *C. greggii*, *C. purpussii*, *C. plicata*, *C. tomentosa*, *Cordia abyssinica*, *C. alba*, *C. alliodora*, *C. axillaris*, *C. boissieri*, *C. brevispicata*, *C. caffra*, *C. cana*, *C. coyucana*, *C. curassavica*, *C. dentata*, *C. dichotoma*, *C. discolor*, *C. ecalyculata*, *C. eleagnoides*, *C. ferruginea*, *C. gerascaulescens*, *C. gharaf*, *C. globosa*, *C. goeldiana*, *C. leucocephala*, *C. macleodii*, *C. magnoliaefolia*,

C. morelosana, C. multispicata, C. myxa, C. nodosa, C. oaxacana, C. podocephala, C. rufescens, C. salvadorensis, C. sebastena, C. sellowiana, C. sonorae, Cordia spp. (12), *C. stellata, C. stenoclada, C. taguahyensis, C. tetandra, C. tinifolia, C. trichotoma, Cryptantha albida, C. angustifolia, C. confertiflora, C. maritima, C. muricata, Cynoglossum amabile, C. furcatus, C. grande, C. occidentale, C. officinale, Echium plantagineum, E. vulgare, Ehretia anacua, E. coerulea, E. cordifolia, E. elliptica, E. laevis, E. taiwaniana, Hackelia floribunda, H. mundula, H. virginiana, Heliophytum* (= *Heliotropium*) *indicum, Heliotropium angiospermum, H. fiarioides, H. fruticosum, H. indicum, H. inundatum, H. limbatum, H. nebonii, H. nelsonii, H. oaxacanum, H. ovalifolium, H. parviflorum, H. procumbens, H. pueblense, Heliotropium spp.* (2), *H. strigosum, H. subulatum, H. supinum, H. ternatum, Lappula mexicana, L. pinetorum, Lithospermum calycosum, L. distichum, L. hirsutum, L. pringlei, L. revolutus, L. strictum, L. viride, Lobostemon fruticosus, L. glaucophyllus, L. montanus, L. oederiaefolius, L. trichotomus, Macromeria discolor, M. exserta, Mertensia ciliata, Mimophytum omphaloides, Moritzia dusenii, Myosotis australis, M. saruwagedica, M. acirpioides, Onosmodium strigosum, O. virginianum, Patagonula americana, Pectocarya spinosa, P. selosa, Plagiobothrys nothofulvis, P. torreyi, Psilolaemus revolutus, Rhabdia* (= *Rotula*) *lycioides, Thaumatocaryon tetraquetrum, Tournefortia bicolor, T. calycina, T. densiflora, T. glabra, T. gnaphaloides, T. hartwegiana, T. hirsutissima, T. mutabilis, T. petiolaris, T. sarmentosa, Tournefortia sp., T. volubilis, Trichodesma angustifolium, T. physaloides, Trichodesma sp., Trigonotis inobleta, T. peduncularis, T. pleiomera, T. procumbens.*

Some taxonomists include *Cordia, Coldenia, Patagonula,* and *Rhabdia* with *Ehretia* in the Ehretiaceae.

BROMELIACEAE
46 genera; 2,110 species

The bromeliads are native to tropical and warm America. Perhaps we are most familiar with them as a source of pineapple, Spanish moss, and many ornamentals. Some species are sources of fiber.

A positive test for alkaloids had been reported for the pineapple, *Ananas comosus*. Tests on 45 samples including 41 species in ten other genera failed to give a positive result. These species were negative: *Aechmea bracteata, A. disticantha, A. ornata, Ananas sativus, Billbergia macrolepsis, Bromelia pinguin, Bromelia sp., Dyckia crocea, D. sellowa, Dyckia sp., Hechtia ghresbreghtii, H. glomerata, H. podantha, H. texana, Pitcairnia karwinskiana, Quesnelia inbricata, Tillandsia achyrostachys, T. andrieuxii, T. benthamiana, T. bulbosa, T. caput-medusae, T. fasciculata, T. ionantha, T. juncea, T. luvida, T. recurvata, T. schiedeana, Tillandsia spp.* (6), *T. tenuifolia, Vriesia carinata, V. friburgensis, V. gladioliffora, V. platynema, Vriesia sp., V. vagans.*

BRUNIACEAE
11 genera; 69 species

The family is South African; some are cultivated for cut flowers. No alkaloids are known. Eight samples representing seven species gave but one positive result, *Berzelia intermedia*. The remainder were negative: *Berzelia abrotanoides, B. lanuginosa, Brunia laevis, B. nodiflora, Nebelia* (= *Brunia*) *paleacea, Staavia radiata*.

BURSERACEAE
8 genera; 540 species

This family occurs in tropical America and in the northeastern portions of Africa. The latter region's species are most familiar as sources of frankincense and myrrh since biblical times. Some have use as ornamentals.

Positive alkaloid tests have been recorded for species of *Commiphora, Boswellia,* and *Protium.* In this study, a total of 149 samples including 95 species gave positive results for *Protium macgregorii* (1/4), *P. neglectum,* and one other undetermined species of that genus.

Negative tests were obtained for the remainder of the samples: *Boswellia serrata, Bursera aptera, Bursera sp.* aff. *aptera, B. arbo-*

rea, B. arida, B. ariensis, B. attenuata, B. bicolor, B. bipinnata, B. citronella, B. confusa, B. copallifera, B. corycensis, B. crenata, B. cuneata, Bursera sp. aff. *cuneata, Bursera sp.* aff. *denticulata, B. diversifolia, B. excelsa, B. fagonoides, B. galeottiana, B. glabrifolia, B. grandifolia, B. graveolens, B. heteresthes, B. hindsiana, B. instabilis, B. jorullensis, B. kerberi, B. lancifolia, B. leptophlocos, B. longipes, B. microphylla, B. morelensis, B. multijuga, B. nesopola, B. occulta, B. odorata, B. palmeri, B. penicillata, B. sarcopoda, B. schlechtendalii, B. simaruba, Bursera sp.* aff. *simaruba, B. submoniliformis, B. terebenthus acuminata, B. tomentosa, B. trifoliata, B. trimera, B. vejar-vasquezii, Canarium acutifolium, C. album, C. australascium, C. australianum, C. maluense, C. pimelum, C. vitiense, Commiphora africana, C. caryaefolia, C. edulis, C. glandulosa, C. harveyi, C. marlothii, C. merkeri, C. mollis, C. neglecta, C. pyracanthoides, C. rehmannii, C. schimperi, C. tenuipetiolata, Elaphrium simarouba, Garuga floribunda, Haplolobus floribundus, H. glandulosus, H. leeifolius, H. robustus, Protium copal, P. guianensis, P. heptaphyllum, P. kleinii, P. nodulosum, P. paraensis, P. polybrotum, Protium spp.* (4), *P. spruceanum, P. tenuifolium, P. unifoliatum, Tetragastris balsamifera, Trattinnickia rhoifolia.*

BUXACEAE
5 genera; 60 species

The family is primarily of the tropics and subtropics of the Old World. *Pachysandra procumbens* of the eastern United States is a common ground cover; *Simmondsia* is the source of a substitute for whale oil. Others are ornamentals.

The "Buxus alkaloids," as the nitrogenous compounds isolated from this family are commonly known, have been recorded from 33 species. The genus *Simmondsia* has been placed in a family of its own, Simmondsiaceae; its seeds have been reported to give a positive alkaloid test but other than a cyanoglycoside, no alkaloid has been isolated.

Two samples were tested: *Buxus lancifolia* was positive, *Sacrococca hookeriana* was not.

C

CABOMBACEAE
2 genera; 8 species

This is a family of aquatics of the tropical and warm temperate areas. They are often used in aquaria. Earlier literature has the genera in the Nymphaceae but Cronquist gives them separate family status.

Alkaloids are not known; *Brasenia schreberi* and *Cabomba caroliniana* tested negative.

CACTACEAE
130 genera; 1,650 species

The Cactaceae are indigenous to the New World. They are economically important as ornamentals; the fruits of *Opuntia* are used as food; the peyote (*Lophophora williamsii*) is a well-known hallucinogen.

Several genera of the family are alkaloidal; the alkaloids are of several types and have been the subject of reviews. Samples (36) covering 33 species were tested to give, as expected, positive tests for *Hylocerus undatus, Lophocerus schottii, Lophophora williamsii, Pachycereus pecten.*

Alkaloids were also detected in the following species: *Cereus sp., Echinocactus enneacanthus, E. roetteri, E. sarissophorus, Opuntia dillenii.*

Negative results were obtained with the following: *Cryptocereus sp., Echinocactus acanthodes, E. horizonthalonius, E. ingens, Echinocereus longisetus, E. mojavensis, E. stamineus, Heliocereus speciosus, Mammilaria dioica, Myrtillocactus geometrizans, Opuntia atrispina, O. auriantiaca, O. cantabrigensis, O. cranacea, O. humifusa, O. imbricata, O. leptocaulis, O. occidentalis, O. tomentosa, Pereskia aculeata, Pereskia sp., Rhipsalis baccifera, R. cassytha, Texocactus melocactiformis.*

CALLITRICHACEAE
1 genus; 17 species

This unigeneric family has been included in others by taxonomists over the years and is now generally accepted as a family of its own. It is of no economic importance.

Little chemical work has been done; alkaloids are not known. Only one species, *Callitriche stagnalis*, gave a positive test (1/2); *C. heterophylla* and *C. verna* were negative.

CALYCANTHACEAE
3 genera; 9 species

This small family is of importance as ornamentals with fragrant flowers. Alkaloids are common in the family.

The known positive species, *Calycanthus floridus* (1/2), *C. glaucus* (2/2), and *C. occidentalis* were also found positive in this study, as was *Chimonanthus praecox*.

CALYCERACEAE
6 genera; 55 species

This small family is related to the Compositae but has none of its economic importance. Alkaloids are not known.

Neither *Acicarpha spathulata* nor *Boöpis bupleuroides* gave positive tests in this study.

CAMPANULACEAE
87 genera; 1,950 species

The family is important chiefly for its large number of ornamentals. It has wide distribution throughout the temperate and subtropical regions with some representatives usually confined to high elevations of the tropics. Some taxonomists have had *Lobelia* in a family of its own (Lobeliaceae), but this genus is now generally included with the other bellflowers. Alkaloids are known, particularly in the genera *Lobelia* and *Campanula*. Lobeline has medicinal use and has served as a substitute for nicotine in attempts to "cure" the nicotine habit.

One hundred and twenty samples of the family were tested representing 96 species; several positive results were obtained from samples known from earlier reports to have been alkaloidal: *Campanula carpathica, C. medium, Isotoma longiflora, Lobelia anceps, L. cardinalis* (2/2), *L. cliffortiana, L. fulgens, L. inflata, L. langeana* (2/2).

Other alkaloid-positive species included the following: *C. collinia, C. glomerata, C. latifolia, C. medium, C. tommasiniana, Centropogon sp., Cyanea angustifolia, Cyphia assimilis, C. bulbosa, C. elata, Hippobroma longifolia* (2/2), *Isotoma petraea, Lobelia* aff. *aguana, L. anceps, L. cardinalis, L. chinensis, L. cliffortiana, L. exaltata, L. decipiens, L. fulgens, L. gruina* (1/2), *L. hassleri* (3/3), *L. laxiflora* (6/10), *L. nicotinaefolia* (2/2), *L. pyramidalis, L. sinaloae, L. syphilitica, Lobelia spp.* (3/6), *L. splendens, L. stenophylla, Phyteuma orbiculare, Siphocampylyus duploserratus, Siphocampylus sp.* (1/2), *S. sulfurens, S. umbellatus, S. verticillatus, Wahlenbergia arenaria, W. banksiana, W. caledonica*.

Negative tests were obtained with the following species: *Campanula americana, C. aucheri, C. barbata, C. caespitosa, C. fenestrellata, C. filicaulis, C. garganica, C. portenschlagiana, C. prenenthoides, C. waldensteinia, Campanumoea* (= *Codonopsis*) *laccinifolia, Centropogon spp.* (2), *C. surinamensis, Clermontia persicifolia, C. kakeana, Codonopsis lancifolia, Cyphia triphylla, Lobelia angulata, L. coerulea, L. coronopifolia, L. ehrenbergii, L. exaltata, L. linearis, L. nuda, L. pinifolia, Lobelia spp.* (3), *L. tomentosa, Phyllocharis* (= *Ruthiella*) *subcordata, Pratia* (= *Lobelia*) *concolor, P. reniformis, Prismatocarpus diffusus, P. pedunculata, P. rogersii, Siphocampylus lycoides, Siphocampylus sp., Triodanis biflora, T. perfoliata, Wahlenbergia androsacea, W. marginata, Wahlenbergia sp., W. undulata*.

CANELLACEAE
5 genera; 16 species

This small family occurs in the tropical regions of the Caribbean, Madagascar, and Africa. The genus *Canella* is valued as an ornamental and by some as a condiment (wild cinnamon).

Chemical investigation of the family has been scant; *Capsicodendron* (= *Cinnamodendron*) *madagascariensis* has yielded a quarternary base.

Six samples representing two species were tested with a positive result obtained for *Capsicodendron dinisii* (1/5). *Warburgia ugandensis* was negative.

CANNACEAE
1 genus; 25 species

The Canna family is primarily one of the New World tropics; according to some taxonomists, three species are indigenous to Asia and Africa. A few are used as ornamentals.

Little is known of the chemistry of this family; alkaloids have not yet been detected. Tests on 13 samples including nine species of *Canna* were without positive result: *Canna coccinea, C. flaccida, C. generalis, C. indica, Canna spp.* (5).

CAPPARIDACEAE
45 genera; 675 species

This family is paleotropic and closely related to the mustards (Cruciferae). One member is familiar as the kitchen spice, capers (*Capparis spinosa*). Others are cultivated as garden ornamentals.

Nitrogen-containing substances are known.

One hundred and nineteen samples including 81 species were tested to give the known positive *Capparis tomentosa* along with a number of other species: *Boscia albitrunca* (1/2), *B. foetida* 1/2), *B. mossambicensis, B. salicifolia* (2/3), *Cadaba aphylla* (1/2), *C. ternitaria* (1/2), *Capparis angustifolia, C. asperifolia, C. brassii, C. erythrocarpa, C. incana* (4/6), *C. indica, C. odoratissima* (1/2), *C. oleoides, C. verrucosa, C. zeylanica, C. zipelliana, Courbonia* (= *Maerua*) *glauca* (2/3), *Crataeva benthamii, C. tapia* (1/4), *Forchhammeria pallida* (1/2), *F. trifoliata, Maerua angolensis, M. caffra, M. silgrii, M. parvifolia, M. pubescens, Thilachium africanum* (1/2).

Some botanists consider the genus *Cleome* in a family of its own, Cleomaceae. Three samples of the genus were positive: *C. diandra* (1/2), *C. hirta*, and *Cleome sp.* (1/6). Other plants included in the Cleomaceae were negative: *Cleome aculeata, C. dendroides, C. gynandra, C. hemsleyana, C. psoralaefolia, C. ruballa, C. serrata, Cleome spp.* (6), *C. speciosissima, C. spinosa, C. suffruticosa, C. viscosa, Cleomella longipes, C. obtusifolia, C. parviflora, C. perennis, Gynandropsis* (= *Cleome*) *pentaphylla, Isomeris* (= *Cleome*) *arborea, Polanisia* (= *Cleome*) *adodecandra, P. icosandra, Polanisia sp., Polanisia tenuifolia, P. uniglandulosa, P. viscosa.*

Alkaloid-negative Capparidaceae included the following: *Boscia rehmanniana, Cadaba kirkii, C. natalensis, Capparis citrifolia, C. coccolobifolia, C. flexuosa, C. lanceolata, C. linesta, C. nummularia, C. rudatsii, Capparis spp.* (3), *C. yco, Crateava palmeri, Forchhammeria sessilifolia, Koeberlinia spinosa, Maerua guerichii, M. juncea, M. parvifolia.*

CAPRIFOLIACEAE
16 genera; 365 species

The honeysuckles are chiefly of the northern hemisphere, notably in Asia and North America. Over 100 species are valued as ornamentals.

The presence of alkaloids in the family was doubted until the isolation of chelidonine from *Symphoricarpos albus*, and an alkaloid from *Lonicera xylostemon*. The phytochemistry of the family was the subject of a thesis some 20 years ago.

The genus *Sambucus* is sometimes placed in a family of its own; it will be included here in the Caprifoliaceae.

Positive tests for alkaloids were obtained from *Viburnum prunifolium* (1/2) (previously known) along with *Abelia floribunda, A. grandiflora* (1/3), *A. spathulata* (1/2), *Lonicera japonica* (1/9), *L. morrowii* (1/2), *L. pilosa* (1/3), *Sambucus pubens, Symphoricarpos microphyllus* (1/4), *Viburnum cassinoides* (1/2), *V. odoratissimum* (2/2), *V. sieboldii, V. suspensum* (4/4), *V. taiwanianum* (2/10), *Weigela hortensis* (1/2).

The following gave negative results: *Abelia coriacea, Alseuosmia quercifolia* (in a separate family, Alseuosmiaceae, by some authorities), *Diervilla sessilifolia, Kolkwitzia amabilis, Leycesteria formosa, Lonicera affinis, L. caprifolium, L. ciliosa, L. confusa, L. gracilis, L. hypoglauca, L. interrupta, L. involucrata, L. mackii, L. sempervirens, Lonicera sp., L. subspicata, L. transarisanensis, L. wilsonii, Pentapyxis* (= *Leycesteria*) *stipulata, Sambucus australis, S. callicarpa, S. canadensis, S. coerulea, S. edulis, S. intermedia, S. javanica, S. mexicana, Symphoricarpos longiflorus, S. orbicularis, S. racemosa, S. vaccinoides, Triosteum perfoliatum, Viburnum acerifolium, V. acutifolium, V. caudatum, V. cotinifolium, V. dentatum, V. dilatatum, V. elatum, V. foetidum, V. juncundum, V. lentago, V. luzonicum, V. macrocephalum, V. microphyllum, V. morrisonense, V. plicatum, V. recognitum, V. recurifolium, V. rufidulum, V. rugosum, V. scabrellum, V. setigerum, V. stellatum, V. tailoense, V. tinus, V. wrightii, Weigela grandiflora.*

CARICACEAE
4 genera; 31 species

The Caricaceae are chiefly tropical American with two genera in tropical Africa. The familiar papaya is a member of this family.

The presence of alkaloids has been recorded and benzylisothiocyanates are known. *Carica papaya* has been reported to give positive alkaloid tests and such was found to be the case here in five of nine samples.

Carica heterophylla, Jacaratia dodecaphylla, J. spinosa, and *Pileus* (= *Jacaratia*) *mexicanus* were negative.

CARPINACEAE
3 genera; 47 species

This is a north temperate zone family, chiefly east Asian. It is usually included in the Betulaceae, to which it is closely related. Alkaloids are not known. In this study, six species of *Carpinus* were alkaloid-negative: *C. caroliniana, C. japonica, C. kawakamii, C. minutiserrata, C. rankanensis, C. tropicalis.*

CARYOCARACEAE
2 genera; 24 species

This is a small family of tropical America. The more important genus, *Caryocar*, is a source of good lumber and its nuts yield a butterlike, edible fixed oil.

No positive alkaloid tests were obtained with samples of *Caryocar brasiliense, C. glabrum, C. microcarpum, C. villosum*.

CARYOPHYLLACEAE
89 genera; 1,070 species

This is essentially a north temperate zone family with a few representatives in south temperate regions and in the tropics at high altitudes. It is known for its many ornamental flowers, including the familiar carnation.

In earlier literature there appear reports of the sporadic occurrence of alkaloids in the family. One hundred and sixteen samples representing 78 species were tested. In agreement with earlier studies, *Dianthus caruthisianorum* and *Lynchnis flor-cuculi* were found to be alkaloid-positive as well as the following species: *Agrostemma walkeri, Corrigiola littoralis* (2/2), *Dianthus allwoodi, D. alpinus, D. armeria, D. arvernensis, D. bolusii, D. deltoides, D. integer, Dianthus sp., Drymaria cordata* (1/5), *Lychnis alpina, Pollichia campestris, Silene lacinata, S. californica*.

Negative tests were obtained with the following samples: *Arenaria bryoides, A. caroliniana, A. decussata, A. lanuginosa, A. lycopodioides, A. macrodenia, A. reptans, Cardionema ramosissima, Cerastium arvense, C. brachypodium, C. cuspidatum, C. gloweratum, C. keysseri, C. papuanum, C. rivulare, Cerastium sp., Corrigiola andina, Cucubalus baccifera, Dianthus armeria, D. basuticus, D. mooiensis, D. namaensis, D. superbus, Kohlrauschia* (= *Petrorhagia*) *prolifera, Lychnis alba, L. alpina, Paronychia brasiliana, P. mexicana, Polycarpea corymbosa, P. eriantha, P. spicata, Polycarpon tetraphyllum, Sagina japonica, S. papuana, S. procumbens, Saponaria ocymoides, S. officinalis, Silene antirrhina, S. capensis, S. caroliniana, S. cucubalus, S. laciniata, S. stellata,*

S. verecunda, Siphonychia diffusa, spergula arvensis, S. bocconii, S. macrotheca, S. marina, Stellaria aquatica, S. arisanensis, S. cuspidata, S. micrantha, S. jamesiana, S. media, S. memorum, S. neglecta, S. ovata, Stellaria sp. (2), S. uliginosa.

CASUARINACEAE
4 genera; 70 species

This is an Australian family furnishing timber and ornamental trees. Little is known of its chemistry; alkaloids have not been detected. All samples tested here were negative: *Casuarina cristata, C. deplancheana, C. equisetifolia, C. glauca, C. lepidophloia, C. littoralis, C. potamophila, Casuarina sp., Gymnostoma papuana.*

CELASTRACEAE
94 genera; 1,300 species

Taxonomists have shifted the genus *Hippocratea* from this family to one of its own and back again. It will be treated here as a member of the Celastraceae, a family of wide distribution except in the arctic regions. Its only economic importance is as a source of ornamentals and of khat (*Catha edulis*), a popular stimulant of the Middle East.

It contains alkaloids among which are the maytansinoids, which have had considerable interest as antitumor agents. Samples (165) of 92 species gave the following as positives previously known: *Catha edulis* (2/3), *Euonymus atropurpureus, Hippocratea indica, Maytenus mossambicensis.*

In addition, the following species were positive: *Bhesa archoboldiana, Cassine crocea, C. kraussiana* (1/3), *Celastrus tetramerus, Euonymus lanceifolia, E. yeddensis* (1/2), *Gymnosporia (= Maytenus) senegalensis* (2/2), *Hartogia (= Hartogiella) capensis* (1/2), *Hippocratea acapulcensis, H. nitida, Loesneriella crenata, Loesneriella macrantha, Maytenus glaucesens, M. guianesis* (2/4), *M. nemorosus, Maytenus sp.* (2/5), *Salacia insignis, Tripterygium regelii* (3/3).

The following were negative: *Acanthothamnus aphyllus, Cassine aethiopica, C. capensis, C. maritimum, C. papillosa, C. pubescens, C. tetragona, Celastrus monospermoides, C. novoguineensis, C. orbicularis, C. pringlei, C. punctatus, C. scandens, Crocoxylon transvaalense, Elaeodendron (= Cassine) curtipendulum, E. capense, E. glaucum, Euonymus acuto-rhombifolia, E. alata, E. bungena, E. echinatus, E. laneum, E. nikoensis, E. oxyphyllus, E. sieboldianus, E. trichocarpus, Euonymus sp., Goupia glabra, Gymnosporia montana (= Maytenus emarginata), Lophopetalum toricellense, Maytenus acuminatus, Maytenus sp.* aff. *rigida, M. alaternoides, M. cymosus, M. ilicifolia, M. oleoides, M. peduncularis, M. phyllanthoides, M. senegalensis, M. undata, M. undulatus, Microtropis fokienensis, Mortonia hidalgensis, M. latisepala, M. palmeri, M. scaberrima, Orthosphenia mexicana, Perrottetia alpestris, P. arisanensis, Perrottetia sp., Plenckia populnea, Plenckia sp., Pseudocassine transvaalensis, Pterocelastrus echinatus, Pterocelastrus sp., Putterlickia verrucosa, Rhacoma scorpia (= Crossopetalum), Salacia erythrocarpa, S. papuana, S. sororia, Schaefferia pilosa, S. stenophylla, Siphonodon celestrineus, S. peltatus, Wimmeria acapulcensis, W. concolor, W. confusa, W. percifolia, Wimmeria sp., Zinowiewia integerrima.*

CENTROLEPIDACEAE
3 genera; 28 species

This small family ranges from southeast Asia to Australia. It is of no known economic importance. Neither alkaloids nor other chemical constituents of the family have been described. *Centrolepis philippensis* was negative for alkaloids.

CEPHALOTAXACEAE
1 genus; 4 species

This unigeneric Asian family is cultivated as an ornamental. Alkaloids are not known but in the present survey *Cephalotaxus fortunei* gave a positive test.

CERATOPHYLACEAE
1 genus; 2 species

This is a family of cosmopolitan aquatic plants often used in aquaria, but in nature it also serves as a shelter for disease-bearing snails and mosquitos. *Ceratophyllum demersum* was alkaloid-negative.

CERCIDIPHYLLACEAE
1 genus; 1 species

Cercidiphyllum japonicum is indigenous to China and Japan and has some importance for its lumber and as an ornamental. Little is known of its chemistry; three samples were negative for alkaloids.

CHENOPODIACEAE
120 genera; 1,300 species

Members of this family are found most often in xerophytic and halophytic habitats and are of worldwide distribution. Familiar foodstuffs (e.g., beets, spinach, Swiss chard) and wormwood (*Chenopodium anthelminiticum*) are members of this family.

Betacyanins and betaxanthins, along with alkaloids in some genera, are found in the family but no one type is prominent or considered characteristic. Of 87 species tested in this survey, the following were confirmed alkaloid-positive as indicated in the earlier work on the family: *Atriplex canescens* (2/8), *A. semibaccata, A. vestita, Beta vulgaris, Chenopodium album* (1/7), *Salsola kali, Suaeda fruiticosa, S. linearis.*

Additional alkaloidal species included the following: *Atriplex polycarpa* (1/2), *A. rosae, Chenopodium ambrosoides* (1/3), *C. nuttaliae* (1/3), *Lophiocarpus burchellii* (now placed in the Phytolaccaceae), *Spirostachys* (= *Heterostachys*) *africanus, Suaeda nigrescens.*

Negative tests were obtained with the following: *Allenrolfea occidentalis, Anthrocnemum africanum, Atriplex acanthocarpa, A. angulata, A. arenaria, A. cineria, A. confertifolia, A. expansa, A. hy-*

menelytra, A. jubata, A. lentiformis, A. limbata, A. linifolia, A. muelleri, A. muricata, A. nummularia, A. obovata, A. patula, A. pentandra, A. serenana, Bassia divaricata, B. hirsuta, B. obliquicuspis, B. paradoxa, Blackiella (= *Atriplex*) *inflata, Chenopodium acuminatum, C. album, C. arizonicum, C. botrys, C. bushianum, C. cahuensis, C. californicum, C. filicifolium, C. foetidum, C. fremonti, C. graveolens, C. missouriensis, C. multifidum, C. murale, C. nitrariaceum, C. rubrum, Chenopodium spp., C. strictum, Cycloloma atriplicifolium, Enchylaena tomentosa, Eurotia lanata, Exomis axyroides, E. microphyllum, Kochia scidarua, Monolepis nuttalliana, Rhagodia baccata, R. linifolia, R. spinescens, R. madaio, Salicornia australis, S. bigelovii, S. europaea, S. glabrescens, S. pacifica, S. virginica, Salsola pestifer, Salsola spp.* (2), *S. subsericea, S. zeyheri, Sacrobatus vermiculatus, Spinacia oleracea, Suaeda californica, S. diffusa, S. mexicana, S. torreyiana.*

CHLORANTHACEAE
4 genera; 56 species

With the exception of a single genus in the New World, this is a family of the tropics and semitropics of the Old World. *Chloranthus glaber* is used as an ornamental in California.

There has been little chemical investigation of the family; a few amides have been characterized in *Chloranthus.*

No alkaloids were detected in *Chloranthus elatior, C. glaber, Hedyosmum artocarpus, H. brasiliense, Hedyosmum spp.* (2).

CISTACEAE
7 genera; 135 species

The Cistaceae are found in the warmer parts of the northern hemisphere, particularly in the Mediterranean region. Their only economic importance is as ornamentals.

The family has had very little chemical investigation; alkaloids have not been found except for a positive test in an unidentified species of *Cistus* obtained in the present survey.

The following species were negative: *Cistus ladeniferus, C. villosus, Halimium exaltatum, Helianthemum corymbosum, H. glomeratum, Helianthemum sp., Hudsonia ericoides, Lechea minor, L. racemosa, L. tripelata, L. villosa.*

CLETHRACEAE
1 genus; 64 species

This unigeneric family is mostly American. A few species are used as ornamentals for their fragrant flowers.

Alkaloids are unknown. Tests on 23 samples including 13 species of *Clethra* were without positive result: *Clethra acuminata, C. alnifolia, C. broadwayana, C. laevigata, C. lanata, C. macrophylla, C. mexicana, C. quercifolia, C. scebra, Clethra spp.* (3), *C. suaveolens.*

COCHLOSPERMACEAE
2 genera; 20-25 species

This small family of tropical distribution is now included by Mabberley in the Bixaceae; *Cochlospermum vitifolium* is used as an ornamental.

The seeds of *Amoreuxia* have been reported to give a positive test for alkaloids; a more recent report has alkaloids in *Cochlospermum planchonii.* In this study, nine samples representing five species gave positive results for *Cochlospermum gillivraei* and *C. vitifolium* (1/5); *Amoreuxia palmatifida, Cochlospermum orinocense,* and *C. religiosum* were negative.

COMBRETACEAE
20 genera; 500 species

This is a pantropical family of little present economic value. *Terminalia catappa* is cultivated for its edible nuts, and a few other genera are ornamentals.

Alkaloids (caffeine, harmans, oxazolidines, pyridines) are known. One hundred and eleven samples encompassing 73 species gave positive tests as follows: *Buchenavia kleinii, B. sericarpa, Bucida buseras, B. macrostachya* (1/2), *Combretum apiculatum* (2/3), *C. caffrum, C. erythrophyllum, C. hereroense* (1/3), *Terminalia amazonica.*

These species were negative: *Anogeissus pendula, A. schimperi, Buchenavia* aff. *sericarpa, B. tomentosa, Cacoucia* (= *Combretum*) *coccinea, Calycopteris floribunda, Calycopteris* (= *Getonia*) *sp., Combretum cacoucia, C. argenteum, C. caffrum, C. calastroides, C. coccineum, C. extensum, C. farinosum, C. fruiticosum, C. gossweileri, C. gueinzii, C. imberbe, C. kraussii, C. laxem, C. mechowianum, C. microphyllum, C. molle, C. mossambicense, C. obovatum, C. ovalifolium, C. paniculatum, C. platypetalum, Combretum spp.* (5), *C. suluense, C. trinitense, C. zeyheri, Conocarpus erectus, Laguncularia racemosa, Lumnitzera littorea, Pteleopsis myrtifolia, Quisqualis indica, Ramatuela sp., Terminalia arjuna, T. australis, T. ballerica, T. catappa, T. dichotoma, T. glabrata, T. guianensis, T. lucida, T. microcarpa, T. mollis, T. muellera, T. myriocarpa, T. obidensis, T. prunioides, T. schumanniana, T. sericea, Terminalia spp.* (3), *T. tomentosa, T. trichopoda, Thioa glaucocarpa.*

COMMELINACEAE
42 genera; 620 species

This family has a wide distribution throughout the tropics and subtropics. It has little economic importance except for a few members cultivated as ornamentals.

Little chemistry of the family is known; alkaloids have not been reported except in an obscure Korean reference to *Commelina communis.*

In this study, 101 samples representing 77 species were tested to give two positive results: *Cyanotis vaga* and *Murdannia semiteres.* The remainder were negative: *Aneilema aequinoctiale, A. angustifolia, A. chiuahuaensis, A. divergens, A. geniculata, A. hockii, A. johnstonii, A. malabaricum, A. nicholsonii, A. plagiocapsa, A. pulchella, Callisia fragrans, Campelia zanonia, Commelina africana, C. aspera, C. benghalensis, C. bracteosa, C. cecilae, C. coe-*

lestis, C. communis, C. cyanea, C. dianthifolia, C. diffusa, C. ecklo-niana, C. ensifolia, Commelina erecta, C. erecta var. *angustifolia, C. forskalaei, C. gerrardi, C. kirkii, C. krebsiana, C. nudiflora, C. obliqua, C. pallida, C. scabra, Commelina spp.* (6), *C. subulata, C. texocana, C. tuberosa, C. umbellata, Cyanotis arachnoides, C. kewensis, C. lanata, C. lapidosa, C. nodiflora, Cymbispatha commelinoides, Dichorisandra hexandra, Floscopa glabrata, F. glomerata, F. scandens, Forrestia chinensis, Gibasis karwinskya-na, G. linearis, G. pulchella, Pollia japonica, Rhoeo* (= *Tradescan-tia*) *discolor, Thyrsanthemum macrophylla, Tinantia erecta, T. fu-gax, T. longipedunculata, T. erecta, Tradescantia crassifolia, T. fluminensis, T. linearis, T. ohiensis, T. virginiana, Tripogandra amplex, T. disgrega, Zebrina* (= *Tradescantia*) *pendula.*

COMPOSITAE
1,314 genera; 21,000 species

The composites can almost compete with the orchids for the title of the largest family of flowering plants. They are found worldwide and in almost all habitats and have economic importance as a source of foods (e.g., lettuce, artichokes), insecticides (pyrethrum), dyes (safflower), folk medicines, and many ornamentals.

Alkaloids are not uncommon in the family but many are of un-known structure. They include amides, found in several genera, which have been considered alkaloids in the broadest sense. Per-haps most familiar of all are the alkaloids of *Senecio* and relatives, which are of importance as stock poisons.

Sesquiterpene lactones are also found throughout the family and some of these may be responsible for reports of positive alkaloid tests, inasmuch as their structural features can give positive reac-tions with the Dragendorff reagent.

The ready availability of members of this large family resulted in a large number of samples for testing and a total of over 2,000 species were examined. In keeping with earlier literature reports, the following were found positive: *Acanthosperum hispidum* (1/4), *Achillea millefolium* (3/2), *Ageratum conyzoides* (1/6), *Ambrosia maritima, Arctium minus, Artemesia tridentata, Baccharis cordifo-*

lia (1/3), *B. halmifolia* (3/4), *Bidens pilosa* (2/16), *Cacalia florida-na* (1/5), *Calendula officinalis* (1/3), *Centaurea cyanus*, *C. maculo-sa* (1/6), *C. melitensis*, *Centratherum muticum* (2/7), *Coreopsis basilis*, *C. lanceolata*, *Cosmos sulphureus*, *Dicoma anomala*, *Emilia sonchifolia* (1/3), *Erechites hieracifolia* (1/4), *Eremanthus sphaerocephala* (1/2), *Eupatorium odoratum*, *E. perfoliatum* (2/7), *E. purpureum*, *E. rotundifolium* (2/2), *E. serotinum*, *Helenium autumnale* (5/8), *Helianthus anuus* (2/3), *Liatris spicata*, *Matricaria chamomilla* (2/4), *Osteospermum spinescens*, *Parthenium hysterophorus*, *Senecio douglasii* (1/2), *S. glabellus* (1/2), *S. graminifolius*, *S. integerrimus* (2/2), *S. jacobaea*, *S. junceus*, *S. pterophorus*, *S. viminalis*, *S. vulgaris*, *Solidago serrata* (2/10), *Tanacetum vulgare* (4/6), *Verbesina enceloides* (6/7), *V. serrata*, *Xanthium pungens* (1/2), *A. strumarium* (1/4).

Positive tests were also obtained for the following: *Acanthocephalus cadamba* (2/2), *Acanthosperum brasilium*, *Acanthospermum sp.*, *Ageratum corymbosum* (3/7), *A. gaumeri*, *A. salicifolium* (1/3), *A. scabrusculum* (1/2), *Ambrosia arborescens*, *A. artemesifolia* (5/6), *A. cumanensis*, *A. hispida* (2/2), *A. peruviana* (1/2), *A. phyllostachys*, *Amellius strigosus*, *Anisopappus africanus*, *Anthemis arvensis* (1/2), *A. cotula* (1/3), *Aplopappus* (= *Haplopappus*) *spinulosus*, *Archebaccharis mucronata* (1/2), *Arctotheca calendula* (1/2), *Arctotis acaulis*, *A. cuprea*, *A. leiocarpa*, *Arnica montata*, *Artemesia afra* (1/2), *A. annua*, *A. californica* (2/3), *A. capillaris* (1/2), *A. douglasiana*, *A. dranunculoides*, *A. indoriciana*, *A. klotzchiana* (1/2), *A. mexicana* (2/4), *Artemisia sp.* (1/2), *Aspilia africana* (2/4), *Aster aethiopicus*, *A. echinatus*, *A. hyssopifolius*, *Aster sp.*, *Athanasia fasciculata*, *A. pinnata*, *A. tomentosa* (2/2), *A. trifurcata*, *Baccharis calvescens*, *B. curitibensis*, *B. elaeagnoides*, *B. eliocla-da*, *B. emoryi* (2/2), *B. glomeruliflora*, *B. megapotamica*, *B. mille-flora*, *B. pilolaris*, *B. platypoda* (2/2), *Baccharis sp.* (1/24), *B. tri-mera*, *B. trinervis* (1/7), *B. vauthieri* (1/2), *B. viminea* (3/3), *Bahia absinthifolia* (5/5), *B. anthemoides* (2/2), *B. schaffneri*, *B. xylopoda* (2/4), *Balduina angustifolia*, *Balsamorhiza deltoides* (1/2), *Barroe-tea sessiliflora*, *Berkheya armata*, *B. ferox*, *B. onoporidifolia*, *B. zeyheri* (1/2), *Berlandiera pumila* (1/2), *Bidens aurea* (3/6), *B. ferulaefolia* (2/5), *B. squarrosa* (1/2), *B. triplinervia* (2/12), *Blainvillea gayana* (2/2), *Borrichia arborescens*, *Brachylaena el-*

liptica, B. transvaalensis, Brachymeris montana, Brickellia califor-nica (2/3), *B. cordifolia, B. coulteri* (1/2), *B. diffusa* (1/3), *B. lacina-ta* (1/2), *B. pendula* (1/3), *B. thyrsiflora, B. tomentella, Cacalia lanceolata, C. sulcata, Calea serrata, Calea sp.* (1/19), *Calendula sp., Callilepis leptophylla, C. salicifolia, Calostephane divaricata* (1/2), *Carphephorus corymbosus, Carphochaete grahami, Cassinia compacta, C. phylicaefolia, C. retorta, C. rhizocephalia, Centipeda orbicularis, Chaenactis douglasii, C. glabruscula* (1/3), *Chondro-phora undata, Chrysopsis sp.* (1/3), *Chrysanthemoides monilifera* (1/2), *Chrysocoma tenuifolia, Chrysoma* (= *Solidago*) *paucifloculo-sa, Chrysothamnus viscidiflorus* (1/2), *C. nauseosus* (4/4), *Chryso-thamnus sp., Cineraria fruticetorum, C. lyrata, Cirsium andersonii, Clibadium sp., Cnicus spp.* (1/9), *Conyza canadensis* (2/2), *C. chi-lensis, C. hochstetterii, C. ivaefolia, C. sophiaefolia* (1/2), *Coreop-sis rhayacophila* (1/2), *Corethrogyne filaginifolia* (1/3), *Cosmos ocellatus, Cotula leptoloba, Crassocephalum mannii, Cyathocline lyrata* (1/2), *Dahlia coccinea* (1/3), *D. scapigeroides, Dicoma ca-pensis, D. gerrardii, Dimorphotheca polyptera, Dispargo ericoides, Doellenergia* (= *Aster*) *reticulata, Dugesia mexicana, Dyssodia ac-erosa* (1/3), *D. seleri* (1/3), *Elephantopus mollis* (2/6), *Encelia fari-nosa* (1/2), *E. californica* (2/2), *Erechites atkinsonia* (1/2), *Erigeron annuus* (1/2), *E. foliosus, E. philadelphicus* (1/4), *E. quercifolius, E. scaposus* (1/4), *Eriophyllum ambiguum, E. confertiflorum, E. multicaule, Erlangea inyangana, Espeletia spp.* (2/3), *Eupato-rium adenophorum* (1/2), *E. africanum* (1/2), *E. aff. havanensis, E. pazcuarense, E. album* (2/2), *E. amplifolium* (2/3), *E. aromati-cum, E. aschenbornianum* (2/9), *E. brevipes* (2/3), *E. calophyllum* (2/4), *E. capillifolium* (1/2), *E. compositifolium* (1/2), *E. crenulatum, E. cuneifolium, E. dubium* (1/2), *E. fistulosum, E. greggii, E. hysso-pifolium* (2/2), *E. irrasum, E. linifolium, E. littorale* (2/2), *E. lingus-trinum* (1/9), *E. marietanum* (1/9), *E. mikanoides, E. nummularia, E. petiolare* (1/4), *E. pygnocephalum* (1/14), *E. recurvans* (2/2), *E. roanensis, E. scorondonioides* (3/6), *E. serratum* (2/3), *E. spina-cifolium, E. tomentallum, E. vautherianum* (1/3), *E. wrightii* (3/4).

Euryops abrotanifolius (1/2), *E. angolensis, E. asparagoides, E. lateriflorus, E. laxus, E. linearis, E. linifolius, E. multifidus* (2/3), *E. pectinatus, E. spathaceus, Euryops sp.* cf. *longipes, Flaveria lineatus, Florestina pedata* (2/3), *Fluorensia cernua* (2/6), *Franser-*

ia (= *Ambrosia*) *bipinnatifida, F. dumosa* (2/2), *F. tenuifolia* (2/3), *Galinsoga parviflora* (1/6), *Gamolepis* (= *Steirodiscus*) *minuta, G. chrysanthemoides, G. euryopoides, G. schinzii, Gazania sp., Geigera africana* (1/4), *G. burkei* (3/6), *Geraea canescens, Gerbera jamesoni, G. plantaginea, Gillardia multiceps, Gochnatia polymorpha* (1/2), *Gongyolepis sp.* (1/2), *Gutierrezia berlandieri* (1/2), *Gynura auriantiaca, Haplopappus ericoides, H. pinifolius, H. teretifolius, H. venetus* (1/2), *Helenium amarum, H. bigelovii, H. flexuosum, H. nudiflorum, H. pinnatifidum* (2/3), *H. quadridentatum, Helianthus angustifolius* (2/2), *H. brasiliensis, H. gracilentus* (2/2), *H. elianthus sp.* (1/3), *Helichrysum lepidissimum, H. odoratissimum* (1/2), *H. squamosum, H. teretifolium, Helipterum canescens, H. polyphyllum, Hertia* (= *Othonna*) *palens, Hippia sp.* indet., *Hirpicium echinus, Hymenoclea salsola, Hymenopappus scabiosaeus* (2/3), *Hymenostephium microcephalum, Inula indica, Iva asperifolia* (1/2), *I. axillaris, I. frutescens* (2/3), *I. xanthifolia, Jungia sellowii, Jurinea cyanoides, Kleinia longiflora* (1/2), *Lasiospermum bipinnatum, Lepidospartum squamatum* (1/2), *Lessingia germanorum* (1/2), *Liatris chapmanii* (2/2), *L. gracilis, L. graminifolia* (2/2), *L. pauciflora, L. tenuifolia, L. secunda, Lopholaena cneorifolia, L. corniifolia* (1/2), *L. disticha, Lychnophora affinis, Macowania sp., M. tenuifolia, Mikania paraensis* (1/2), *Mikania sp.* (1/16), *Mollera angolensis, Montanoa grandiflora* (4/5), *M. tomentosa* (1/6), *Moscharia sp.* indet., *Nidorella auriculata, Nolletia arenosa, N. ciliaris, Olearia heterocarpa, Ophryosporus ovatifolius* (1/2), *Osmiopsis astericoides, Osteospermum microcarpum, O. microphyllum, Othonna amplexicaulis* (1/2), *O. carnosa* (1/3), *O. coronopifolia, O. triplinervia, Oxylobus arbutifolius* (1/3), *Palafoxia linearis, P. texana, P. tomentosum, Pegolettia baccharidifolia, P. retrofracta* (2/3), *P. senegalensis, Pentzia globifera, P. incana, P. argentea, P. dentata, P. punctata, P. spinescens, P. virides* (1/2), *Perezia cubatensis, P. microcephalia, P. wrightii* (1/3), *Perityle emoryi, P. microglossa, Perymenium berlandieri, Peyrousea umbellatus, Pheobanthus tenuifolia, Phymaspermum parvifolium, Phymaspermum sp., Piptocarpha leprosa, Pluchea camphorata* (2/3), *P. rosea* (2/2), *Podachaenium eminens, Polymnia laevigata* (1/4), *P. sylphioides, Polypteris* (= *Palafoxia*) *integrifolia, Propteopsis argentea, Psilostrophe gnaphaloides* (1/3), *Pteronia incana, P. sor-*

dida, P. spinescens, Ratibida pungens (1/2), *Rudbeckia mollis, Sa-bazia humilis* (1/2), *Sabazia sp., Saussurea discolor, S. formosana, Schistostephium sp., Schkuhria pinnata* (1/2), *Sclerocarpus frutes-cens* (1/2), *S. spathulatus.*

The genus *Senecio* is well known for its alkaloids; the following have been known to be positive: *Senecio* aff. *deformis, S. albonervis, S. arenius, S. aronicoides, S. aureus* (2/2), *S. bracteatus, S. burchellii* (1/2), *S. calcarius* (1/3), *S. cineraroides, S. consanguineus, S. coronatus, S. deformis* (2/2), *S. filifolius, S. fraudulentus, S. glastifolius, S. grandiflorus, S. harveianus, S. hastatus, S. incomptus, S. longiflorus, S. longifolius, S. monoensis, S. nivens, S. obovatus, S. pentactinus, S. pleistocephalus, S. pohlii, S. prenanthoides* (5/5), *S. procumbens, S. pterocaulis, S. purpureus, S. rhyncholaenus, S. salingus* (3/8), *S. sanguisorbe, Senecio spp.* (7/18), *S. strictifolius, S. vulneraria.*

The following were alkaloid-positive in the present survey: *Solidago altissima* (1/3), *S. boottii, S. graminifolia* (3/3), *S. leavenworthii, S. leptocephala, S. microcephala* (3/6), *S. odora* (1/2), *S. sp.* indet., *S. stricta, Sonchus elliotianus, Spilanthes ocynifolia* (1/2), *Stephanomeria pauciflora* (1/2), *S. tenuifolia, Stevia connata* (1/3), *S. elliptica, S. elongata* (1/6), *S. micrantha, S. menthaefolila, S. monardaefolia* (4/11), *S. nepetaefolia* (2/9), *S. purpurea* (2/6), *S. reglensis* (2/4), *S. rhombifolia* (2/16), *S. acabrella, S. trifida* (1/2), *Stoebe alopecuroides, S. capitata, S. plumosa, S. vulgaris, Symphyopappus compressus, Tagetes sp., Tetradymia axillaris* (1/2), *Tetragonotheca helianthoides* (1/4), *Tithonia brachypappa* (1/2), *Townsendia mexicana* (2/3), *Trichogonia hirtiflora, Tricholepsis claberrima* (1/2), *Tridax brachylepis, T. coronopifolia* (1/3), *T. procumbens* (1/11), *T. trifida, Trixis angustifolia* (1/5), *T. longifolia* (1/4), *T. radialis* (1/4), *T. verbasciformis* (1/2), *Ursinia anethoides, U. filiformis, U. annua, Venegasia carpesoides* (1/2), *V. glabrata, V. intermissa, Venegazia sp.* (3/6), *V. stricta* (1/2), *V. virginica, Vernonia angustsifolia* (1/5), *V. colorata, V. corymbosa* (1/2), *V. crassa* (1/2), *V. elongata* (2/2), *V. fastigata* (1/2), *V. gigantea, V. lampropappa, V. leptolepsis, V. monocephala, V. noveborazencis, V. nudiflora* (2/3), *V. palmeri, V. poskeana, Vernonia sp.* (3/30), *V. schwenkiaeifolia, Virguiera angustifolia, V. bicolor, V. brevifolia* (1/3), *V. deltoides* (1/2), *Virguiera sp.* (2/6), *V. multiflo-*

ra, V. trachyphylla (1/2), *Wedelia menotriche, Wyethia angustifolia, Xanthium italicum* (1/3), *X. pennsylvanicum* (2/2), *Zaluzania angusta* (2/2), *Z. globosa, Z. mollissima* (2/2), *Z. montagnaefolia* (2/5), *Z. robinsonia* (1/3), *Z. triloba* (2/3), *Zexmenia lantanifolia* (1/3), *Zinnia peruviana.*

Negative tests were obtained with the following: *Acanthospermum australe, Achaetogeron ascendens, Achillea borealis, Achyrocline alata, A. saturoides, Achrocline spp.* (2), *Actinomeris alternifolia, A. tetraptera, Adenocaulon bicolor, Adenopappus persicaefolium, Adenostemma brasilianum, A. caffrum, A. viscosum, Aganippea* (= *Jaegeria*) *bellidiflora, Ageratum candidum, A. classocarpum, A. houstonianum, A. littorale, A. longifolium, Ageratum spp.* (2), *A. tomentosum, Agiabampoa congesta, Agrianthus empetrifolium, Albertinia brasiliensis, Aldama dentata, Alomia wendlandii, Ambrosia confertiflora, A. polystachyia, A. trifida, Ammobium alata, Anacyclus depressus, Anaphalis contorta, A. lorentzii, A. morrisonicola, A. nagasawai, A. margaritea, A. subumbellatum, Anaxeton asperum, Anisocoma acaulis, Anisopappus canescens, A. dentatus, A. lastii, Antennaria neodioica, A. plantaginifolia, A. rosea, Anthemis aizoon, A. hausknechtii, Aphanostephus humilis, Aplopappus* (= *Haplopappus*) *hartwegi, A. spinosus, A. venetus, Aplostephium* (= *Haplostephium*) *lasserinoides, Aracium* (= *Crepis*) *tolucanum, A. asperifolia, A. glandulosa, A. hieracioides, Archibaccharis hirtella, A. mucronata, A. sescenticeps, A. androgyna, Arctium lappa, Arctotis candida, A. laevis, A. petiolata, A. stachadifolia, Arnica chamissonis, A. cordifolia, Artemesia annua, A. axillaris, Aspilia brachyphylla, A. caruthii, A. absinthum, A. australis, A. balchandrum, A. dracunculus, A. dubia, A. ludoviciana, A. pycnocephala, A. rothrockii, Artemesia sp., A. stellariana, A. vulgaris, A. foliacea, A. foliosa, A. helianthoides, A. laevissima, A. linearifolia, A. kotschyi, A. montevidensis, A. procumbens, A. schimperi, A. setosa, Aspilia spp.* (2), *A. verbenoides, Aster ageratoides, A. alpinus, A. bakerianus, A. carolianus, A. cordifolius, A. decumbens, A. divaricatus, A. ericaefolius, A. exilis, A. filifolius, A. gracilis, A. gymnocephalus, A. haplopappus, A. hirsuticaulis, A. junceus, A. lima, A. linearifolius, A. luteus, A. macrophyllus, A. montevidensis, A. muricatus, A. nova-angliae, A. patens, A. peglerae, A. pilosus, A. prenanthoides, A. puniceus,*

A. reflexus, A. rotundifolius, A. schlechteri, A. schreberi, A. scopulorum, A. simplex, A. spectabilis, Aster spp. (2), *A. spinosus, A. subulatus, A. taiwanesis, A. tenacetifolius, A. tenifolius, A. umbellatus, A. walteri, Anthanasia acerosa, A. dentata, A. parviflora, Athrixia elata, A. heterophylla, A. phylicolides, A. rosmarinifolia, Atractylis lancea, Atrichoseris platyphylla, Baccharidastrum* (= *Conyza*) *triplinervium, Baccharis angustifolia, B. angusticeps, B. anomala, B. articulata, B. axilaris, B. calvescens, B. camporum, B. cassinifolia, B. conferta, B. douglasii, B. dracuncufolia, B. dracemna, B. elliptica, B. erigeroides, B. gaudichaundiana, B. glutinosa, B. helichysoides, B. heterophylla, B. ilinita, B. lateralis, B. lingustrina, B. macrocephala, B. nucinella, B. orgyalis, B. pentaptera, B. plummerae, B. potosina, B. puberula, B. ramulosa, B. ramiflora, B. sarthoides, B. sebastianopolitana, B. serraefolia, B. serrulata, B. sessiliflora, B. soralescens, Baccharis spp.* (22), *B. subspanthulata, B. tarchonanthoides, B. thesioides, B. trimera, B. trinervis, B. vaccinoides, B. varians, B. vauthieri, B. vernonoides, B. weirii, Baeria* (= *Lasthenia*) *chrysostoma, Baileya multiradiata, B. thurberi, Balduina uniflora, Balsamorhiza sagittata, Barroetea setosa, B. subligera, Berkheya barbata, Berkheya sp.* aff. *carlinopsis, B. decurrens, B. echinaceae, B. fruticosa, B. heterophylla, B. insignis, B. pinnatifolia, B. radula, B. seminivea, B. setifera, B. speciosa, Berkheya sp., B. spinosimna, Berlandiera lyrata, Bidens anthemoides, B. anthriscoides, B. bidentoides, B. bigelovii, B. bipinnata, B. biternata, B. cernua, B. comosa, B. coronata, B. cynapiifolia,B. discoides, B. gardneri, B. graveolens, B. insecta, B. mitis, B. osthrutioides, B. polylepis, B. racemosa, B. rubicundula, B. segetum, B. serrulata, B. schaffneri, B. schimperi, B. serrulata, Bidens spp.* (5), *B. vulgata, Blanchetia heteroticha, Blepharipappus scaber, Blumea aurita, B. balsamifera, B. glomerata, B. jacquemontii, B. lacera, B. lacinata, B. myriocephala, B. pubigera, B. riparia, B. spectabilis, B. virens, Borrichia frutescens, Brachyglottis repandra, Brachylaena discolor, B. neriifolia, B. rotundata, Brachymeris* (= *Phymaspermum*) *bolusii, Brasilia sickii, Brickellia cavanillesii, B. conduplicata, B. corymbosa, B. glutinosa, B. nelsonii, B. nutanticeps, B. nutans, B. odontophylla, B. pacayensis, B. palmeri, B. paniculata, B. pringlei, B. pulcherrima, B. scoparia, Brickellia spp.* (3), *B. spinulosa, B. squarrosa, B. verbenacea, B. veronicaefolia, Cacalia* (= *Arno-*

glossum) ampullacea, C. decomposita, C. sessiliflora, Calea acaulis,
C. ferruginea, C. hispida, C. hymenolepis, C. integrifolia,
C. oxylepis, C. pachyphylla, Calea sp. aff. *pachyphylla, C. parreifolia, C. parvifolia, C. peduncularis, C. pinnatifida, C. pringlei,*
C. rotundifolia, C. rupestris, C. scabrifolia, C. solidaginea, Calea
spp. (18), *C. tenuiifolia, C. tridactyla, C. urticifolia, C. zacatechichi, Calotis lappulacea, C. latiuscula, Calyptocarpus vialis, Carduus acanthoides, C. tenuiflorus, Carminatia tenuiflora, Carpesium*
acutum, Carphochaete grahami, Carthamus tinctorius, Cassinia
tenuifolia, C. laevis, C. leptophylla, C. quinquefaria, C. spectabilis,
C. vauvilliersii, Castalis nudicaulis, Celmisia incana, C. spectabilis, Cenia (= Lancisia) pectinata, C. turbinata, Centaurea solstitialis, Centipeda minima, Centratherum punctatum, Chaenactis fremontii, Chaetopappa modesta, Chaptalia exscapa, C. nutans,
C. runcinata, Chrysactinia acerosa, C. mexicana, Chrysanthemum
carnosulum, C. frutescens, C. indicum, C. leucanthemum, C. segetum, C. sinense, Chrysophisis aff. *graminifolia, C. decumbens,*
C. flexuosa, C. mariana, C. mixta, Chrysopsis sp., C. villosa, Chuquiraga sp., Cichorium intybus, Cineraria erosa, Cineraria sp.,
Cirsium arvense, Cirsium californicum, C. coulteri, C. discolor,
C. horridulum, C. japonicum, C. kawakamii, C. lanceolatum, Cirsium sp., C. vulgare, Clibadium armani, C. asperum, C. pueblanum, Clibadium spp. (2), *C. sylvestre, C. surinamensis, Cnicus*
benedictus, C. arvensis, C. nivalis, C. pinetorum, Cnicus spp. (8),
Conyza aegyptica, C. ambigua, C. bonariensis, C. chilensis, C. erythrolaena, C. gnaphalioides, Conyza spp. (3), *C. ulmifolia, Coreocarpus arizonicus, Coreopsis californica, C. gigantea, C. heterogyna, C. lucida, C. mutica, C. petrophiloides, C. pringlei, Coreopsis*
spp. (3), *C. tripteris, Corymbium villosum, Cosmos acabiosioides,*
C. bipinnatus, C. caudatus, C. crithmifolius, C. montanus, C. parviflorus, C. scabiosioides, Cotula coronopifolia, Craspedia globosum, C. uniflora, Crassocephalum crepidioides, C. picridifolia,
C. sarcobasis, C. mannii, Crepis acuminata, C. capillaris, C. japonica, Crossostephium chinese, Cullumia ciliaris, Cynara scolymus,
Cynthia (= Krigia) virginica, Dahlia pinnata, D. latifolium, D. tomentosum, D. velutina, Dasphyllum sp., Denekia capensis, Desmanthodium lanceolatum, Dichaeta (= Schaetzellia) haenkeana,
Dichrocephala latifolia, D. bicolor, D. integrifolia, Dicoma argyro-

phylla, D. galpinii, D. sessiliflora, D. zeyheri, Dicoria canescens, Dicranocarpus parviflorus, Dimerostema lipioides, Dimerostema spp. (4), *Dimorphotheca sinuata, D. cuneata, Diplostephium sp., Dubautia plantaginea, Dyscritothamus filifolius, Dyssodia cancellata, D. chrysanthemoides, D. gnaphalopsis, D. neaei, D. pentachaeta, D. pinnata, D. polychaetum, D. porophylla, D. setifolia, D. tapetiflora, Echinops echinatus, Eclipta alba, E. prostrata, Elephantopus angustifolius, E. elatus, E. elongatus, E. pilosus, E. scaber, Elephantopus sp., E. spicatus, Elytropappus rhinocerotis, Emilia flammea, E. pagittata, Encelia albescens, Epaltes gariepina, Erechites arguta, Erechites spp.* (3), *E. valerianifolia, Eremanthus crotonoides, E. elaeagnus, E. eriopus, E. glomeratus, Eremanthus spp.* (3), *Erigeron aphanactis, E. bonariensis, E. breweri, E. calcicola, E. canadensis, E. clokeyi, E. compositus, E. delphinifolius, E. divergens, E. filifolius, E. floribundus, E. glaucus, E. inornatus, E. macrosatus, E. maximus, E. mucronatus, E. ochroleucus, E. peregrinus, E. pumilus, E. repens, Erigeron spp.* (5), *E. tweedii, Eriophyllum lanatum, Erlangea laxa, E. sessifolia, E. tomentosa, E. trifoliata, Eroeda (= Oedera) capensis, E. imbricata, Erythrocephalum sp., Espeletia corymbosa, E. hartwegiana, Espeletia sp., Ethulia conyzoides, Eupatorium adamantinum, Eupatorium sp.* aff. *adenospermum, Eupatorium sp.* aff. *areolare,* aff. *betonicaefolium,* aff. *espinosarum,* aff. *oresbinum, Eupatorium ambylolepsis, E. areolare, E. ascendens, E. barbacense, E. bellidifolium, E. betonicaeforme, E. bigelovii, E. bupleurifolium, E. calminthaefolium, E. chapalense, E. chinese, E. colestinum, E. collinum, E. congestum, E. conspicuum, E. crassirameum, E. daleoides, E. deltoideum, E. edmundoi, E. ellipticum, E. espinosarum, E. formosanum, E. gaudichiaudianum, E. glabratum, E. glandulosum, E. gracillum, E. greggii, E. gracillicante, E. havanense, E. halmifolium, E. haenkeanum, E. hebebotryum, E. hidalgense, E. indulaefolium, E. incarnatum, E. intermedium, E. itatyaiense, E. ivaefolium, E. japonicum, E. karwinskianum, E. laeve, E. laetevireus, E. laevigatum, E. lanigerum, E. leptophyllum, E. leucoderme, E. liebmannii, E. littorale, E. linifolium, E. longipes, E. lundeanum, E. lucidum, E. macrostemon, E. macrocephalum, E. macrophyllum, E. maximiliani, E. mendezii, E. mohrii, E. montanum, E. monilifolium, E. morifolium, E. molissimum, E. muelleri, E. multifidum, E. multiflosculosum,*

E. neaeanum, *E. nelsonii*, *E. oligocephalum*, *E. ortegae*, *E. organe-*
sis, *E. ovaliflorum*, *E. palmare*, *E. pazcuarense*, *E. pilosum*, *E. porri-*
ginosum, *E. polycephalum*, *E. prunellaefolium*, *E. pycnocephalum*,
E. pulchellum, *E. pumilum*, *E. quadrangulare*, *E. rhomboideum*,
E. riparium, *E. rugosum*, *E. saggitiflorum*, *F. semiserratum*, *Eupa-*
torium spp. (40), *E. spinacifolium*, *E. spinosarum*, *E. spathulatum*,
E. subpenninervium, *E. subintegrum*, *E. tashiroi*, *E. tetragonium*,
E. thyrosoideum, *E. tremulum*, *E. vernicosum*, *Euryops athanasiae*,
E. brevipappus, *E. oligoglossus*, *E. rupestris*, *E. tenuissimus*,
E. virgineus, *Eutetras palmeri*, *Filago californica*, *Flaveria angus-*
tifolia, *F. anomala*, *F. australasica*, *F. bidentis*, *F. repanda*, *Flaveria*
sp., *F. trinervia*, *Fleischmannia arguta*, *Flourensia resinosa*, *Fran-*
seria (= *Ambrosia*) *acanthocarpa*, *F. confertiflora*, *F. malvaceae*,
Gaillardia aestivalis, *G. pulchella*, *G. tonkiev*, *Galinsoga ciliata*,
Gamochaeta (= *Gnaphalium*) *spicata*, *Gamolepis* (= *Steirodiscus*)
chrysanthemoides, *G. trifurcata*, *G. brachypoda*, *Garuleum album*,
G. pinnatifidum, *Gazania linearis*, *G. krepsiana*, *G. pygmaea*,
G. rigida, *Geigeria schinzii*, *Gerbera glandulosa*, *G. aspitiflora*,
G. crocea, *G. discolor*, *G. natalensis*, *G. piloselloides*, *Gibbaria*
ilicifolia, *G. scabra*, *Gnaphalium* aff. *brevicaspa*, *G. attenuatum*,
G. beneolens, *G. bicolor*, *G. californicum*, *G. chartaleum*, *G. chi-*
lense, *G. diaicum*, *G. hypoleucum*, *G. indicum*, *G. involucratum*,
G. japonicum, *G. lavendulifolium*, *G. leptophyllum*, *G. luteo-album*,
G. microcephalum, *G. morii*, *G. multiceps*, *G. obtusifolium*, *G. occi-*
dentalis, *G. oxyphyllum*, *G. peregrinum*, *G. purpurescens*, *G. pur-*
pureum, *G. ramosissimum*, *G. rhodantum*, *Gnaphalium spp.* (12),
G. spicatum, *Gochnatia hypoleuca*, *Gongylolepis martiniana*, *Gor-*
ceixia sp., *Gorteria corymbosa*, *Grangea maderaspatana*, *Grinde-*
lia gladulosa, *G. inuloides*, *Guardiola angustifolia*, *G. mexicana*,
Guizotia abyssinica, *G. scabra*, *Gutenbergia gossweileri*, *Guten-*
bergia sp., *Gutierrezia grandis*, *G. microcephala*, *G. sarothrae*,
Gymnosperma glutinosa, *Gynura angulosa*, *G. crepioides*, *G. di-*
varicata, *G. flava*, *G. formosana*, *G. segetum*, *G. vibellina*, *Halo-*
carpha lyrata, *H. scaposa*, *Haplopappus acaulis*, *H. cooperi*,
H. divaricatus, *H. linearifolius*, *H. palmeri*, *H. squarrosus*, *H. stolo-*
niferus, *Haplostephium janarinoides*, *Hecubaea* (= *Helenium*) *scor-*
zonerifolia, *Hedypnois cretica*, *Helenium amphibolum*, *H. mexica-*
num, *H. microcephalum*, *H. scorzoneraefolium*, *H. decapetalus*,

H. debilis, H. grosseratus, H. radula, Helianthus spp. (2), *H. tephrodes, H. tuberosus, Helichrysum aculatum, H. adscendens, H. allioides, H. appendiculatum, H. argyrophyllum, H. argyrosphaerum, H. athrixifolium, H. aureonitens, H. crispum, H. cylindricum, H. caespititum, H. cerastioides, H. confertifolium, H. coriaceum, H. crispum, H. cymosum, H. decorum, H. ericaefolium, H. felinum, H. foetidum, H. kraussii, H. kirkii, H. lancifolium, H. latifolium, H. lucilioides, H. miconiaefolium, H. mundtii, H. nudiflorum, H. nitens, H. orbiculare, H. panduratum, H. paniculatum, H. pentzioides, H. petiolatum, H. platypterum, H. rugulosum, H. sesamoides, H. setosum, Helichrysum spp.* (3), *H. splendidum, H. stenopterum, H. swynnertonii, H. thapsus, H. umbraculigerum, H. undatum, H. vestitum, H. zeyheri, Heliopsis annua, H. helianthoides, H. longipes, H. procumbens, Helipterum gnaphaloides, H. speciosissimum, Helminthia* (= *Picris*) *echioides, Hemizonia corymbosa, H. kelloggii, H. multicaulis, Hertia* (= *Othonna*) *alata, Heterolepis saliena, Heterosperma pinnatum, Heterothalamus sp., Heterotheca chrysopsidis, H. grandiflora, H. inuloides, H. subaxillaris, Hidalgoa ternata, Hieracium abscissum, H. comaticeps, H. crespidispermum, H. flagellare, H. florentinum, H. greenei, H. gronovii, H. leucotrichium, H. pratense, Hieracium sp., Hippia frutescens, H. pilosa, Hirpicium bechuanese, H. gracilis, H. integrifolium, Hofmeisteria pluriseta, Hulsea vestita, Hymenostephium* (= *Viguiera*) *cordatum, Hypericophyllum angolense, Hypochoeris alata, H. radicata, Hypochoeris sp., Ichthyothere latifolia, Ichthyothere sp., Ifloga aristulata, I. reflexa, Inula cappa, I. glomerata, I. helenium, I. paniculata, I. royleana, Inula sp., Inulopsis* (= *Podocoma*) *seaposa, Iostephane heterophylla, Ischnea elachoglossa, Isocarpha oppositifolia, Isostigma sp., I. speciosum, Iva ciliata, Ixeris japonica, I. microcephala, Ixiolaena brevicompta, Jaegeria hirta, Jaegeria sp., Jaumea peduncularis, Jungia floribunda, Kanimia* (= *Mikania*) *oblongifolia, K. nitida, Keysseria gibbsiae, K. radicans, Krigia virginica, Kuhnia* (= *Brickellia*) *rosmarinifolia, Lachnospermum ericifolium, Lactuca capensis, L. floridana, L. graminifolia, L. intybacea, L. canadensis, L. biennis, L. debilis, L. sativa, L. scariola, Lactuca sp., Lagascea angustifolia, L. decipiens, L. glandulosa, L. helianthifolia, L. heteropappus, L. rubra, Lagenphora stipitata, Laggera* (= *Blumea*) *alata, Launaea asplenifolia, Layia glandulosa,*

Leucactinia bracteata, Leucopholis capitata, Leysera gnaphaloides, L. tenella, Liabum adenotrichum, L. deppanum, L. discolor, L. glabrum, Liatris elegans, L. laevigata, Ligularia stenocephala, L. tussilaginea, Lipochaeta integrifolia, Lophalaena platyphylla, Loxothysanus pedunculatus, L. sinuatus, Lychnophora affinis, L. bunioides, L. damazioi, L. martiniana, L. pinifolia, L. rosmarinifolia, L. slavioides, Lychnophora spp. (6), *L. tomentosa, L. trichocarpha, Lygodesmia aphylla, Machaeranthera tortifolia, Madia elegans, M. sativa, Mallinoa* (= *Ageratina*) *corymbosa, Matricaria matricarioides, Melampodium montanum, M. oblongifolium, M. perfoliatum, Melampodium divaricatum, M. montanum, M. oblongifolium, M. gracilis, Melanthera nivea, M. scandens, Metalasia cephalotes, M. gnaphalodes, M. muricata, M. seriphiifolia, Microglossa pyrifolia, Microglossa sp., M. volubilis, Microseris heterocarpus, M. lindleyi, Microspermum mummulariaefolium, Microspermum sp., Mikania cordata, M. guaco, M. hatschbachii, M. burchellii, M. cipoensis, M. cordata, M. decumbens, M. diversifolia, M. lasiandra, M. matthewsii, M. micrantha, M. neurocaula, M. oblongifolia, M. obtusifolia, M. obtusata, M. officinalis, M. parvifolia, M. ternata, M. trinervis, M. scabra, M. scandens, M. sericea, M. sessilifolia, Mikania spp.* (16), *M. subverticillata, M. ternata, M. vitifolia, Milleria quinquefolia, Minuria denticulata, M. leptophylla, Montanoa frutescens, M. mollissima, M. myriocephala, M. patens, M. pauciflora, M. speciosa, Moquinia paniculata, Mutisia campanulata, Myriactis humilis, M. longipedunculata, Myriocephalus stuartii, Nestlera* (= *Relhania*) *conferta, N. humilis, Neurolaena lobata, N. macrocephala, Nicolletia edwardsii, N. occidentalis, Nidorella microcephala, N. resedifolia, N. undulata, N. welwitchii, Nolletia rarifolia, Notoptera* (= *Otopappus*) *epalacea, Oldenburgia arbuscula, O. asterotricha, O. paradoxa, Olearia albida, O. arborescens, O. elliptica, O. ferresii, O. furfuracea, O. nernstii, O. nummularifolia, O. ramulosa, O. rana, Oligandra imbricata, Ophryosporus petraceus, Osmites* (= *Relhania*) *hirsuta, Osteospermum calendulaceum, O. ciliatum, O. clandestinum, O. corymbosum, O. ecklonis, O. imbricatum, O. monocephalum, O. muricatum, O. scariosum, O. sinuatum, Osteospermum spp.* (2), *O. spinosum, Othonna auriculaefolia, O. natalensis, O. parviflora, Othonna sp., Oxylobus adscendens, Palafoxia linearis, P. leucophylla,*

P. texana, Parthenium argentatum, P. bipinnatifolium, P. integrifolium, P. incanum, Pectis angustifolium, P. arenaria, P. dichotoma, P. fasciculiflora, P. febrifuga, P. latisquama, P. stenophylla, P. elongata, Pegolettia lanceolata, Pentzia globosa, P. lanata, P. sphaerocephala, P. suffruticosa, Perezia adnata, P. alamani, P. cuernavacana, P. dugesii, P. hebeclada, P. lozani, P. michoacana, P. purpusii, P. reticulata, P. rigida, Perezia sp., Perymenium acuminatum, P. blepharolepis, P. buphthalmoides, P. cervantesi, P. glandulosum, P. gracile, P. mendezii, P. pringlei, P. rude, P. subsquarrosum, P. tenellum, P. verbesenoides, Petasites sp., Phaenocoma prolifera, Picris hieracioides, Pinaropappus roseus, Piptocarpha angustifolia, P. axillaris, P. cinearia, P. oblonga, P. rotundifolia, Piptocarpha sp., Piqueria pilosa, P. trinervia, Pleiotaxis amoena, P. pulcherima, Pluchea adnata, P. dioscorides, P. foetida, P. indica, P. longifolia, P. odorata, P. purpurescens, P. rubelliflora, Pluchea sp., Podosperma (= Podotheca) angustifolium, Pollalesta milleri, Polymnia maculata, P. riparia, P. uvedalia, Porophyllum amplexicaule, P. calcicola, P. coloratum, P. ellipticum, P. fracile, P. loneare, P. macrocephalum, P. pantatum, P. scoparium, Porophyllum spp. (2), P. tagetoides, P. viridiflorum, Prenanthes altissima, Printzia polifolia, Psathyrotes annua, P. ramosissima, Pseudoelephantopus spicatus, Pseudobaccharis (= Baccharis) ligustrina, Psiadia arabica, Psilactis (= Machaeranthera) breviligulata, Pterocaulon alopecuroideum, P. angustifolium, P. serrulatum, Pterocaulon spp. (2), P. undulatum, Pteronia camphorata, P. divaricata, P. glauca, P. pallens, P. paniculata, Pulicaria angustifolia, Pyrethrum cineariifolium, Pyrrhopappus georgianus, Rafinesquia californica, R. mexicana, Ratibida columnaris, P. pinnata, Relhania genistaefolia, R. squarrosa, R. trinervia, Rhysolepsis mordensis, Riencourtia oblongifolia, Rudbeckia fulgida, R. hirta, R. lacinata, R. triloba, Rumfordia alcorte, R. floribunda, Sabazia liebmannii, S. michoacana, Salmea oligocephala, Santolina chamaecyparissus, Sanvitalia procumbens, Sanvitalia sp., Sartwellia humilis, Saussurea affinis, S. carthamnoides, S. deltoidea, Schistocarpha bicolor, S. oppositifolia, Schistostephium artemesifolium, S. crataegifolium, S. heptalobum, S. oxylobum, S. rotundifolium, Schistostephium sp., Schkuhria anthemoidea, Schlechtendalia glandulosa, S. luzulaefolia,

Sclerocarpus schiedeanus, Sclerocarpus sp., S. uniserialis, Sclerolepsis uniflora, Selloa plantaginea.

In spite of the well-known ability of *Senecio* species to produce alkaloids, many of the species collected and tested in this survey gave negative results: *Senecio achilleaefolius, S. absolutescens, S. alvarezensis, S. amplifolius, S. andrieuxii, S. angulifolius, S. angustifolius, S. argutus, S. aschenbornianus, S. barba-johannis, S. bidwillii, S. brachycodon, S. brachypodus, S. brasiliensis, S. carnosus, S. chapalensis, S. chicarrensis, S. colensoensis, S. cordifolius, S. deltoides, S. elegans, S. erosus, S. ellipticus, S. erubescens, S. freemanii, S. galpinii, S. grossidens, S. guadalajarensis, S. hartwegi, S. hemmensdorfii, S. heracleifolium, S. ilicifolius, S. inornatus, S. insularis, S. isatideoides, S. junodii, S. lalienus, S. latifolius, S. leptoschizus, S. leptostachium, S. limosus, S. maritimus, S. microglossus, S. monoensis, S. oleosus, S. obtusiflorus, S. pandurifolius, S. paulensis, S. peltiferus, S. petasites, S. pinnilatus, S. platanifolius, S. praecox, S. pubigerus, S. quinquelobus, S. ramantaceus, S. radulaefolium, S. repandus, S. reticulatus, S. rigidus, S. roldana, S. roseus, S. scandens, S. scleratus, S. serra, S. sinuatus, Senecio spp.* (11), *S. stoechadiformis, S. suffultus, S. taiwanensis, S. tamoides, S. tolucanus, S. trixoides, S. triactinus, S. umbellatus, S. verdoorniae, S. vestitus, S. westermanii.*

Other negative species included: *Sericocarpus (= Aster) asteroides, S. tortifolius, Seris (= Gochnatia) amplexifolia, Siegesbeckia orientalis, Silphium compositum, S. perfoliatum, Silphium sp., Simsia adenophora, S. calva, S. foetida, S. lagascaeformis, S. sanguinea, S. tenuis, Solidago bicolor, S. caesia, S. canadensis, S. confinis, S. fistulosa, S. flexicaulis, S. hispida, S. juncea, S. mexicana, S. memoralis, S. paniculata, S. puberula, S. sempervirens, S. virgoaurea, Sonchus asper, S. oleraceus, S. elliotianus, Sonchus spp.* (2), *Sphaeranthus hirtus, S. hispidus, Spilanthes acmella, S. americana, S. grandiflora, S. mauritania, S. uliginosa, Stenachaenium macrocephalum, Stenocline chionaea, Stephanomeria virgata, Stevia aschenborniana, S. berlandieri, S. clinophodioides, S. cordifolia, S. cupatoria, S. elatior, S. eupatoria, S. glandulosa, S. hirsuta, S. jorullensis, S. labifolia, S. lucida, S. micrantha, S. nervosa, S. origanoides, S. paniculata, S. pubescens, S. purpusii, S. pyrolaefolia, S. salicifolia, Stevia spp.* (3), *S. subpubescens, S. tomentosa,*

S. vernicosa, S. viscida, Stifftia uniflora, Stilpnopappus ferrugineus, Symphyopappus cuneatus, S. lymansmithii, S. reticulatus, Symphyopappus spp. (2), *Syncephalantha* (= *Dyssodia*) *decipiens, Synedrella nudiflora, Synedrella sp., Tagetes tenuifolia, T. erecta, T. filifolia, T. florida, T. foetidissima, T. coronopifolia, T. lucida, T. micrantha, T. minuta, T. nelsonii, T. patula, T. peduncularis, T. triradiata, Tagetes sp., T. stenophylla, T. subvillosa, Taraxacum dens-leonis, T. formosanum, T. officinale, Tarchonanthus camphoratus, T. galpinii, Tetradymia canescens, T. glabrata, T. stenolepis, Tetramolopium alinae, T. flaccidum, T. macrum, T. pumilum, Tetraneuris angustata, Tithonia calva, T. diversifolia, T. tagatiflora, T. tubaeformis, Townsendia montana, Tragopogon sp., Trichocline incana, T. polymorphia, Trichocline spp.* (2), *Trichogonia podocarpa, T. villosa, Trichoptilium incisum, Tridax balbisioides, T. palmeri, Tridax sp., T. trilobata, Trigonospermum floribundum, T. melampodioides, Trigonospermum sp., Trilisa odoratissima, T. paniculata, Trixis antimenorrhea, T. angustifolia, T. californica, T. eryngioides, T. mexicana, T. praestans, T. pringlei, T. petiocaulis, Trixis sp., T. verbasciformis, Tussilago farfara, Ursinia eckloniana, U. anethifolia, U. anethemoides, U. chrysanthemoides, U. crithmifolia, U. dentata, Ursinopsis* (= *Ursinia*) *caledonica, Vanillosmopsis spp.* (2), *Venedium* (= *Arctotis*) *fastuosum, V. decerrens, V. semipapposum, Verbesina croceata, V. gracilepes, V. greenmani, V. hypomaleca, V. klattii, V. liebmannii, V. longipes, V. montanoifolia, V. occidentalis, V. oligantha, V. oncophora, V. pedunculosa, V. persicifolia, V. potosina, V. robinsonii, V. sordenses, Verbesina spp.* (7), *V. sphaerocephala, V. stenophylla, V. tetraptera, V. virgata, V. wrightii, Vernonia* aff. *alamani, V. ampla, V. amygdalina, V. anagallis, V. anthelmintica, V. squarrosa, V. arborea, V. arctioides, V. argenta, V. argyrophylla, V. aurantiaca, V. balansae, V. brevirolia, V. capraefolia, V. cephalotes, V. cinerea, V. cistifolia, V. cognata, V. colorata, V. coriacea, V. deppeana, V. diffusa, V. echitifolia, V. eriolepis, V. ervendbergii, V. fruticosula, V. gerberiformis, V. glabrata, V. gracilis, V. grandiflora, V. gratiosa, V. hirsuta, V. hoffmanniana, V. holstii, V. hypochaeris, V. lancibracteata, V. leucocalyx, V. lindbergii, V. luteoalbina, V. mariana, V. megapotamica, V. melleri, V. mucronulata, V. multiflora, V. natalensis, V. niktidula, V. oligocephala, V. oproprdioides, V. ovalifolia, V. pallens, V. patula, V. per-*

rottetti, V. polyanthes, V. polysphaera, V. psilophylla, V. puberula, V. purpurea, V. quinqueflora, V. rubricaulis, V. salicifolia, V. saxicola, V. scorpioides, V. serratuloides, V. shirensis, V. sinclairi, Vernonia spp. (28), *V. staehelinoides, V. steetziana, V. thomasiana, V. tomatella, V. trocediana, V. umbricata, V. vepretorum, V. vestita, V. viscidula, V. warmingii, Viguiera buddleiaeformis, V. dentata, V. eusifolia, V. excelsa, Viguiera sp.* aff. *greggii, V. laciniata, V. linearis, V. pringlei, V. sessilifolia, Viguiera spp.* (4), *V. sphaerocephala, V. tenuis, V. trichophylla, Vittadinia brachycomoides, V. triloba, Volutaria divaricata, Wedelia biflora, W. hispada, W. macrodonta, W. paludosa, Wedelia spp.* (2), *W. stirlingi, W. uniflora, Wunderlichia sp., Xanthium spinosum, Xanthocephalum alamani, X. humilis, Xanthocephalum sp., X. tomentellum, Youngia japonica, Zaluzania cinerascens, Z. aluzania sp., Zexmenia aurea, Z. brevifolia, Z. ceanothifolia, Z. crocca, Z. frutescens, Z. fruticosa, Z. ghiesbrechti, Z. gnaphaloides, Z. helianthoides, Z. michoacana, Zinnia acerosa, Z. angustifolia, Z. anomala, Z. citrea, Z. elegans, Z. greggii, Z. juniperifolia, Z. littoralis, Z. multiflora, Z. pauciflora, Z. pumila.*

CONNARACEAE
20 genera; 380 species

This is chiefly a tropical family of twining shrubs with no great importance other than the use of a few species as fiber or medicinal/ toxic preparations at local levels.

The chemistry of the family is not known; alkaloids have not been reported beyond their presence in *Connarus* and *Cnestis*.

Seven samples representing seven species were tested without positive result: *Byrsocarpus orientalis, Connarus conchocarpus, Connarus sp., Rourea glabra, R. microphylla, Rourea sp., R. surinamensis.*

CONVOLVULACEAE
58 genera; 1,650 species

The morning glories constitute a tropical and subtropical family with some extension into temperate zones. The familiar sweet potato is a member of this family, which includes many ornamentals.

Alkaloids are found in the family, some in the seeds of morning glories, which is responsible for their use as hallucinogens. However, in the study reported here, seeds were not tested. Nonetheless, species known to be positive from earlier reports were found so as follows: *Ipomoea alba* (1/3), *I. tricolor, I. violacea, Quamoclit pinnata, Turbinia* (= *Rivea*) *corymbosa* (1/7).

Other positive tests include the following: *Convolvulus aschersonii, C. capensis, C. multifidus, C. repens, Cuscuta americana,* sometimes assigned to a separate family, *Cuscutaceae, C. potosina* (1/3), *Dichondra carolinensis, Exogonium* (= *Ipomoea*) *bracteatum* (1/3), *Exogonium sp., Ipomoea sp., Porana elutina* (1/2), *Prevostea africana.*

The following species were negative: *Astripomoea malvacea, Bonamia media, Calystegia marginata, Convolvulus arvensis, C. aridus, C. farinosus, C. macrophyllus, C. occidentalis, C. occelatus, C. sepium, Convolvulus sp., Cressa truxillensis, Cuscuta australis, C. gracillima, C. gronovii, C. mitraeformis, C. natalensis, C. racemosa, C. subinclusa, Cuscuta sp., Dichondra argentea, D. repens, Evolvulus martii, E. nivens, E. tenuis, Exogonium conzattii, Falkia repens, Ipomoea adenioides, I. asarifolia, I. albivenia, I. aquatica, I. arachnosperma, I. asterophora, I. batatas, I. bathycolpus, I. blepharophylla, I. bombycina, I. calobra, I. carnea, I. castellata, I. caudata, I. chloroneura, I. clavata, I. coscinosperma, I. crassipes, I. ficifolia, I. gracilisepala, I. heptaphylla, I. heterophylla, I. hochstetteri, I. intrapilosa, I. involucrata, I. leucanthemum, I. littoralis, I. longiflora, I. longifolia, I. lozani, I. magnusiana, I. muelleri, I. muricoides, I. obscura, I. ommanneni, I. palmata, I. paulistana, I. pedatisecta, I. pellita, I. pentaphylla, I. pes-caprae, I. phyllomega, I. plebia, I. polyanthes, I. procumbens, I. purga, I. purpurea, I. quinquefolia, I. ramosissima, I. reniformis, I. sinensis, Ipomoea spp.* (11), *I. stans, I. stolonifera, I. tiliacea, I. tricolor, I. tryrianthina, I. tuba, I. tuberosa, I. verbascoides, I. wolcottiana, I. wrightia, Iseia sp., Jacquemontia blanchettii, J. holoricea, J. nodiflora, J. pentantha, J. perryana, J. sandwicensis, J. tamnifolia, J. violacea, Maripa sp., M. tenuis, Merremia aegyptia, M. dissecta, M. emarginata, M. gemella, M. kentrocaulos, M. macrocalyx, M. peltata, M. pinnata, M. quinata, Merremia spp.* (2), *M. tridentata, M. umbellata, Operculina aegyptia, O. alata, O. alatipes, O. convolvulus, Operculina sp., O. tuberosa, O. turpethum, Porana paniculata, Prevostea* (= *Calycobolus*) *spectabilis, Prevostea sp., Quamoclit coccinea, Q. peduncula-*

ris, Quamoclit sp., Rivea sp., Seddera sp., Turbinia corymbosa, T. holubii, T. oblongata, T. oenotheroides, T. shirensis, T. suffruticosa.

CORIARIACEAE
1 genus; 5 species

This unigeneric family is widely distributed; some are used for ornamentals and the fruits of some are poisonous.

Only one species, *Coriaria myrtifolia*, has been reported to give a test for alkaloids; in this study six samples representing four species of the genus were negative: *C. japonica, C. papuana, C. ruscifolia, C. thymifolia.*

CORNACEAE
12 genera; 90 species

These few genera found mostly in temperate North America and Asia are used chiefly as ornamentals.

The genus *Cornus* has been reported to contain alkaloids. In this study, a total of 39 samples of 32 species gave but one positive test, *Aucuba japonica* (1/3).

Earlier, this family had been split into a number of smaller families, all samples of which were negative for alkaloids: *Curtisia dentata* (Curtisiaceae), *Griscelina littoralis, G. lucida, G. ruscifolia* (Griscelinaceae), *Helwingia laponica* (Helwingiaceae), *Nyssa biflora* (now in Nyssaceae). *Macrocarpium* has been assigned to *Cornus*; *M. officinale* was alkaloid-negative, as were the following species of *Cornus*: *C. alternifolia, C. amomum, C. californica, C. disciflora, C. excelsa, C. florida, C. koosa, C. nutallii, C. obliqua, C. officinalis, C. racemosa, C. sessilis, C. stolonifera, C. stricta.*

CORYLACEAE
35 genera; 1,500 species

This north or south temperate mountain family Hegnauer places in the Betulaceae, Willis in the Corylaceae; Cronquist equates the two.

The chemistry of the family is unknown; no positive alkaloid tests were obtained for *Corylus americana, C. avellana, C. cornuta, C. rostrata, C. sieboldiana,* and *Corynocarpus laevigatus,* which is sometimes placed in a family of its own, Corynocarpaceae, a position not of general agreement. This species had been reported earlier to give a positive test for alkaloids.

CRASSULACEAE
35 genera; 1,500 species

The family is widely distributed but only a few members occur in South America and almost none in Australia and Oceania. They have limited use as ornamentals.

Alkaloids have been found in a few species. In this survey, 81 samples of 64 species were tested to give, as the only positive, the known *Bryophyllum daigremontianum. Crassula expansa, C. maritima* (1/2), *Crassula sp.* cf. *corymbosa, C. vagitana* (1/3), *Echeveria pubescens, Sedum oxypetalum,* and *Sedum sp.* (1/5), which were not known, were also positive.

Negative tests were obtained for *Bryophyllum pinnatum, Cotyledon decussata, C. leucophylla, C. orgiculata, C. paniculata, C. ramosissium, C. wallichii, C. wickensii, Crassula acinaciformis, C. acutifolia, C. alsinoides, C. arborescens, C. argentea, C. argyrophylla, C. cephalophora, C. ciliata, C. falcata, C. lycopodioides, C. mesembryanthemoides, C. multicava, C. nodulosa, C. parvisepala, C. rubicunda, C. rupestris, C. southii, C. tetragona, C. thorncroftii, Dudleya farinosa, D. lanceolata, D. saxosa, Echeveria carnicolor, E. coccinea, C. fulganes, E. glauca, E. nuda, Kalanchoe laciniata, K. longiflora, K. paniculata, K. pinnata, K. rotundifolia, Kalanchoe spp.* (2), *K. spathulata, K. thyrsiflora,* (= *Crassula*) *subulata, Sedum bulbiferum, S. dendraideum, S. hemsleyanum, S. liebmannianum, S. minimum, S. moranense, S. obtusatum, S. sarmentosum, S. telephium, S. ternatum.*

CRUCIFERAE
90 genera; 3,000 species

The mustard family is primarily of the cool areas of the northern hemisphere and is important for the number of food crops it yields:

cabbage, cauliflower, turnips, mustards, horseradish, etc. Some are used as ornamentals.

So-called proto- and pseudoalkaloids are known throughout the family and several reviews of these have appeared. Some of the species were assayed here and likewise found to be alkaloid-positive: *Capsella bursa-pastoris* (1/8), *Lepidium cartilacineum, Nasturtium montanum* (1/2), *Raphanus sativus* (2/3).

Other positive samples included *Alyssum maritinum* (1/2), *Brassica sp.* (1/2), *Cakile edentula* (1/2), *Cardamine lyrata, C. oligosperma, Draba jarullensis, D. mexicana, Draba sp., Eruca sativa* (1/2), *Erucastrum griquense, Erysimum suffrutescens, Heliphila erithimifolia, Lepidium fremontii, L. hyssopifolium, Lesquerella purpurea* (1/2), *Thelypodium pallidum, Thlaspi occitanicum, T. rotundifolium.*

Negative tests were obtained with the following species: *Alliaria officinalis, Arrabidopsis thaliana, Arabis alpina, A. dignia, A. glabra, A. holboellii, A. inoyensis, A. laevigata, A. lyrata, A. maxima, A. muralis, A. sparsiflora, Asta schaffneri, Barbarea vulgaris, Berteroa incana, Brachycarpea varians, Brassica campestris, B. geniculata, B. oleracea, Cakile maritima, Cardamine bulbosa, C. flexuosa, C. hirsuta, Cardamine spp.* (2), *Cardaria draba, Caulanthus amplexicaulis, C. pilosus, Cibotarium divaricatum, Coronopus integrifolius, Descurainia pinnata, D. sophila, Draba confusa, D. myosotioides, D. nivicola, D. pumila, Erysimum argillosum, E. asperum, E. perenne, Erysimum sp., Halimolobus palmeri, Isatis glauca, Lepidium campestre, L. howei-insulae, L. medium, Lepidium spp.* (3), *L. virginicum, Lesquerella sp., Matthiola sp., Nasturtium officinale, Nerisyrenia camporum, N. gracilis, Raphanus raphanistrum, Rapistrum rigosum, Rorippa atrovirens, R. curvisiliqua, Rorippa islandica, R. nasturtium, R. wateri, Sisybrium capense, S. irio, S. linearifolium, S. officinale, Streptanthella longirostris, Streptanthus tortuosus, Synthlipsis greggii, Thelypodium laciniatum, T. integrifolium, T. micranthum, Thysanocarpus curvipes, Warea amplexifolia.*

CRYPTERONIACEAE
5 genera; 11 species

The family is found in Asia, South Africa, and South America. Alkaloids have not been found.

Samples of the 11 species were alkaloid-negative: *Crypteronia cumingii, C. glabrata, C. griffithii, C. lauraefolia, C. laxa, C. leptostachys, C. paniculata, C. pubescens, Heteropyxis natalensis, Henslowia spp.* (2). *Heteropyxis* is sometimes placed in Heteropyxaceae.

CUCURBITACEAE
121 genera; 760 species

This is primarily a tropical and semitropical family with some members in the northern and southern hemispheres. It is perhaps best known as a source of foods: pumpkins, squash, cucumbers, melons, etc.

Alkaloids (pyrazoline derivatives) and some others of undetermined structure have been reported from several genera. The family is known chemically for the cucurbitacins–toxic steroidal substances. In this study, 106 samples in 73 species were tested; two were already known to contain alkaloids, *Momordica charantia* (2/11) and *M. foetida* (1/2). Other positives included: *Coccinea addensis* (1/3), *Cucumis anguria, C. zeyheri, Cucurbita foetidissima* (2/3), *Echinocystis macrocarpa, Marah macrocarpa, Melothria cordata, Momordica cissoides, M. repens* (1/2), *Peponium mackenii, Trichosanthes sp.*

Negative were the following: *Anguria* (= *Citrullus*) *grandiflora, A. triphylla, Benincasia cerifera, B. hispida, Bryonia dioica, Cayaponia racemosa, Cayaponia sp., Citrullus lanatus, C. vulgaris, Coccinea quiqueloba, C. rehmannii, C. sessilifolia, C. variifolia, Ctenolepis cerasiformis, Cucumis hirsuta, C. metuliferus, C. myriocarpus, C. sativus, C. umbrosis, Cucurbitopsis sp., Cyclanthera eremocarpa, C. naudiniana, Cyclanthera sp., Ecballium elaterium, Echinocystis milleflora, E. rongispina, Elaterium cartaginense, Gurania sp., G. spinulosa, Kedrostis sp., Lagenaria mascarena, Lagenaria sp., Luffa acutangula, L. cylindrica, L. operculata, Melothria heterophylla, M. maderaspatana, Melothria spp.* (2), *Momordica clematidea, M. dioica, M. runsoria, Neoalsomitra sarcophylla, Raphanocarpus* (= *Momordica*) *boivinii, Schizocarpum filiforme, Schizocarpum sp., Sechiopsis triqueter, Sechium edule, Sicyos angulatus, S. microphylla, Sicyos sp., Sphaerosicyos* (= *Lagenaria*) *sphaericus, Trichosanthes palmata, T. homophylla, T. dioica, T. kirilowii, Trochomeria hookeri, T. macrocarpa.*

CUNONIACEAE
24 genera; 340 species

Related to the Saxifragaceae, this family is confined almost exclusively to the southern hemisphere. It is of little economic importance; some species are used as ornamentals, and one as timber in New Zealand.

Alkaloids have been reported present in the family. Thirty samples representing 25 species tested in this study failed to give a positive result: *Belangera* (= *Lamanonia*) *speciosa, Ceratopetalum succirubrum, Cunonia capensis, Cunonia sp., Geissois sp., G. montana, Gillbeea papuana, Lamanonia speciosa, L. ternata, Pancheria spp.* (4), *Platylophus trifoliatus, Pullea stozeri, Schizomeria ovata, Spiraenthemum spp.* (3), *Weinmannia paullinifolia, W. ledermannii, W. racemosa, W. silvicola, Weinmannia spp.* (2).

CUPRESSACEAE
17 genera; 113 species

This is a cosmopolitan family with several members important as timber, gums, and resins as well as ornamentals.

There have been a couple of reports of the presence of alkaloids in the family, but these have not been characterized nor do they make a significant contribution to the otherwise terpenoid chemistry of the family.

In this study, the following gave positive alkaloid tests: *Chamaecyparis formosensis* (2/2), *Juniperus virginianum* (1/3), *Thuja occidentalis.*

Negative tests were obtained with the following: *Callitris endlicheri, Chamaecyparis obtusa, C. pisifera, Chamaecyparis sp., C. thyoides, Cupressus benthamii, C. forbesii, C. pygmaea, C. macrocarpa, Juniperus bermudiana, J. californica, J. deppeana, J. flaccida, J. fruticetis, J. monosperma, J. monosperma* var. *graci-*

*lis, J. monticola, J. morrisonicola, J. silicicola, Libocedrus austro-
caledonica, L. bidwillii, L. decurrens, L. formosana, L. papuanus,
Papuacedrus papuanus, Thuja orientalis, Widringtonia cupres-
soides.*

CYCADACEAE
1 genus; 20 species

The unigeneric cycad family is found from East Africa to Japan
and Australia. Other genera, formerly placed in this family, are now
included in the Zamiaceae (q.v.) by some botanists: *Bowenia, Encephalartos, Lepidozamia, Macrozamia,* and *Zamia. Stangeria* has
been separated by some authors into the Stangeriaceae.

The pith of the sago "palm," *Cycas circinalis,* is used as food in
India after preparation to remove toxic alkaloidal constituents that
occur in this genus and in the Zamiaceae.

Samples of *Cycas circinalis, C. revoluta,* and *C. taiwaniana* were
alkaloid-negative.

CYCLANTHACEAE
11 genera; 190 species

A family of herbs, shrubs, and lianas of the West Indies and
South America, some of these plants furnish thatch and brooms; the
fiber of one species is used for making Panama hats.

Alkaloids are not known in the family; two species of *Carludovica* were tested with negative result: *C. palmata, Carludovica sp.* aff.
atrovirens.

CYPERACEAE
115 genera; 1,600 species

The sedges have a worldwide distribution, chiefly in the subarc-
tic and temperate zones of both the northern and southern hemispheres. They are closely allied to the grasses. *Cyperus papyrus* is

used as an ornamental aquatic and was employed in ancient Egypt in "paper" making.

Alkaloids are known in the family, especially in the genus *Carex*. In this study, no alkaloids were detected in 196 samples representing 142 species. The following were tested: *Bulbostylis barbata, B. capillaris, Calyptocarya longifolia, Carex abrupta, C. apressa, C. athrostachya, C. baccans, C. brunnea, C. chrysolepis, C. clavata, C. dietrichiae, C. doniana, C. filicina, C. flava, C. lasiocarpa, C. lurida, C. meadii, C. molestra, C. pocilliformis, C. phacota, C. pocillifolia, C. radiata, C. remotiflora, C. remotispicula, C. rostrata, C. shimada, Carex sp., C. subtransversa, C. tatsutakensis, Carpha alpina, C. glomerata, Chrysithrix capensis, Cladium anceps, C. meyeri, Cyperus alternifolius, C. amabilis, C. andreanus, C. aschenbornianus, C. compressa, C. difformis, C. diffusus, C. esculentus, C. flavus, C. iria, C. malaccensis, C. monophyllus, C. mundulus, C. odoratus, C. papyrus, C. pedunculatus, C. pilosus, C. procerus, C. pumila, C. rotundus, C. sanguinolentus, C. spectabilis, Cyperus spp. (2), C. tenullus, C. unioloides, C. vegetus, Dichromena colorata, Eleocharis acicularis, E. avenicola, E. dulcis, E. limosa, E. tenuis, Ficinia bracteata, F. dichotoma, F. filliformis, F. fusca, F. ferruginea, F. ixioides, F. ramosissima, F. truncata, Fimbristylis complanata, F. dichotoma, F. falcata, F. miliacea, F. monostachya, F. polytrichoides, Fimbristylis sp., F. spicata, F. polytrichoides, F. selicea, Fuirena sp., Gahnia gahniaeformis, G. gaudichaudii, G. xanthocarpa, Heliocharis sp., Hypolytrum latifolium, H. nemorum, H. schraderianum, Kyllingia brevifolia, Lagenocarpus rigidus, Lipocarpha chinensis, Mapania macrocephala, Mariscus cyperinus, M. haematedes, M. riparius, M. sieberianus, Oreobolus ambiguus, Pleurostachys gaudichiaudii, Pycerus polystchyus, P. globosus, P. globosus var. erecta, P. sanguinolentus, Remirea maritima, Rhynchospora globosa, Rhynchospora spp. (2), R. ruba, R. capitellata, Schoenus moschalinus, Scirpus americanus, S. atrovirens, S. californicus, S. crassiusculus, S. erectus, S. microcarpus, S. morrisonensis, S. morrisonicola, S. mucronatus, S. nodosus, S. olneyi, S. platycarpus, S. robustus, S. rubicosus, S. rubrofinatus, Scirpus sp., S. subterminalis, S. validus, S. wallichii, Scleria bebecarpa, S. polycarpa, S. secans, S. sphacelata, Tetraria cuspida-*

ta, T. bromoides, T. ustulata, Thoracostachyum sumatrum, Uncinia australis.

CYRILLACEAE
3 genera; 14 species

This is an American family distributed from southeastern United States and Cuba to Brazil and Colombia. Two species are cultivated as ornamentals.

The chemistry of the family is unknown; *Cliftonia monophylla, Cyrilla racemiflora,* and *C. parvifolia* were alkaloid-negative.

D

DAPHNIPHYLLACEAE
1 genus; 10 species

Daphniphyllum is distributed through eastern Asia from China south through Malaysia to tropical Australia. Some species are used as ornamentals; the Ainu of Japan use the leaves of one species as tobacco.

The genus contains alkaloids, according to earlier reports. Of seven species, including 14 samples tested here, three were positive: *D. gracile, D. membranaceum,* and the previously known *D. calycinum.* Four species were negative: *D. glabrescens, D. glaucens, D. papuanum,* and one unidentified species.

DATICACEAE
3 genera; 4 species

These are trees that range from Malaysia to Australia and in western North America from California to Mexico.

Alkaloids have not been detected in the family; one sample of *Octomelis sumatrana* tested in this study was negative.

DIASPENSIACEAE
5 genera; 13 species

This is a New World family found in cool to arctic regions of the northern hemisphere.

Some are occasionally cultivated as ornamentals. Alkaloids are not known.

Three samples (one each of three species) were tested to give a positive result for *Shortia exappendicula* and negative results for *Pyxidanthera barbulata* and *Shortia transalpina*.

DICHAPETALACEAE
3 genera; 125 species

These tropical trees, shrubs, and lianas are known for their ability to accumulate fluorine in the form of fluoroacetic acid. They are toxic to stock and some are actually cultivated for poisons used on pest animals in Africa. Alkaloids of the pyridine type have been recorded in the family.

Seven samples that included four species were tested in this survey. One of three samples of *Dichapetalum timoriense* was positive. *D. vestitum, Tapura guianensis,* and *T. singularis* were negative.

DILLENIACEAE
12 genera; 300 species

This is a family of the warm and tropical zones, especially those of Australia. Some members are used as timbers, others yield edible fruits, and a few are considered ornamentals.

Caffeine and "some alkaloid-like substances" have been noted in the family. Forty-six samples representing 34 species were tested without positive result: *Curatella americana, Davilla aspera, D. elliptica, D. kunthii, D. lucida, D. rugosa, Davilla sp., Dillenia alata, D. montana, D. papuana, D. philippinensis, Doliocarpus sp., D. sellowianus, Hibbertia aspera, H. candicans, H. glaberrima, H. melhanoides, H. sericea, Hibbertia spp.* (3), *Tetracera nordtiana, T. scandens, T. sellowiana, Wormia* (= *Dillenia*) *biflora, Wormia sp.*

DIOSCOREACEAE
8 genera; 630 species

The yams constitute a tropical and warm-temperate family known as a source of food and for the sapogenins from which steroid hormones are synthesized.

Alkaloids are known in a few members of the family. Tests on 45 samples which included 37 species resulted in three positives in plants known to be alkaloidal (*Dioscorea dumetorum, D. hispida, D. alata*) and in two of 13 other unidentified *Dioscorea* species. The remainder were negative: *Dioscorea abyssinica, D. alata, D. batatas, D. buchananii, D. bulbifera, D. composita, D. cotinifolia, D. doryophora, D. dregeana, D. esculenta, D. hemicrypta, D. hirtiflora, D. mexicana, D. pentaphylla, D. quartiniana, D. retusa, D. sylvatica, D. trifida, Tamus edulis.*

DIPSACACEAE
8 genera; 250 species

The Mediterranean basin and neighboring Eurasia and Africa are the areas of distribution of this family. Some are ornamental, a few are medicinal in Asia, and a couple of species furnish the teasel used in dressing cloth.

Alkaloids of the iridoid type are found in the family. Of eight samples tested in this survey, *Cephalaria attenuata* (2/2), *Dipsacus fullonum,* and *D. sylvestris* were positive; *Cephalaria cephalobotrys* and three species of *Scabriosa* were negative: *S. africana, S. albanesis, S. columbaria.*

DIPTEROCARPACEAE
16 genera; 530 species

The Dipterocarpaceae constitute a tropical family found especially in Malaysia. It yields timber, resins, and an edible fat from one genus (*Shorea*).

The presence of alkaloids has been recorded for *Marquesia*. Ten samples representing six species of other genera were negative in this survey: *Anisoptera kostermansii, A. polyandra, Dipterocarpus angolensis, Monotes adenophyllus, M. autennei, M. glaber.*

DROSERACEAE
4 genera; 85 species

This is a cosmopolitan family of insectivorous plants known as sundews (*Drosera*).

Alkaloids have not been recorded in the family. Seven species of *Drosera* were tested to yield one positive result, *D. spathulata*; *D. auriculata, D. capensis, D. neocaledonica, D. peltata*, and two undetermined species were negative.

E

EBENACEAE
2 genera; 485 species

With major representation in tropical and warm zones and a few temperate species, the Ebenaceae are known for timbers (ebony) and fruits (persimmons).

There have been occasional reports of the presence of alkaloids. In this study, testing of 105 samples in 69 species yielded only one positive result, *Euclea polyandra.*

The following were negative: *Diospyros affinis, D. austro-africana, D. batocana, D. dichrophylla, D. digyna, D. discolor, D. ebenaster, D. ebenum, D. embryopteris, D. eriantha, D. ferrea, D. galpinii, D. glabra, D. guiannensis, D. hebecarpa, D. hillebrandii, D. ierensis, D. inconstans, D. kanjilalii, D. kirkii, D. lycioides, D. mabola, D. melinoni, D. mespiliformis, D. morrisiana, D. nummularia, D. palmeri, D. paniculata, D. papuana, D. peekelii, D. peregrina, D. quitoensis, D. rotundifolia, D. scabrida, D. simii, Dioscorea spp. (8), D. subrotata, D. tomentosa, D. undabunda,*

D. vaccinioides, D. villosa, D. virginiana, D. virgata, D. whyteana, Euclea acutifolia, E. crispa, E. daphnoides, E. divinorum, E. lanceolata, E. natalensis, E. polyandra, E. pseudebenus, E. racemosa, E. schimperi, Euclea sp., E. tomentosa, E. undulata, Maba (= *Diospyros*) *hemicycloides, M. inconstans, Royena* (= *Diospyros*) *lycioides.*

ELAEAGNACEAE
3 genera; 45 species

The family ranges through the warm temperate regions of the northern hemisphere to the tropics, mostly in southern Asia, Europe, and North America. Some members are cultivated as ornamentals.

Alkaloids, including ß-carbolines, are found in the three genera of the family. In the present survey, they were detected in *Eleagnus latifolia*, previously known to contain alkaloids, as well as in *E. macrophylla, E. multiflora,* and *E. pungens* (1/6), while *E. umbellata, E. wilsonii,* and *Shepherdia arguta* were negative.

ELAEOCARPACEAE
10 genera; 520 species

This is a family of the warm and tropical regions excepting the continent of Africa. It has some local uses for timber and fruit, and a few species are considered ornamental.

Hegnauer puts the family in the Tiliaceae. The Elaeocarpaceae are known to contain alkaloids in several genera. Of 47 samples tested, the following species previously reported to contain alkaloids were also found positive here: *Elaeocarpus densiflorus, E. dolichostylus, E. polydactylus.* Seven other species were also determined to be positive: *Aceratium megalosporum, Aristotelia australascia, A. serrata, Elaeocarpus altisectus, E. archboldianus,* an unidentified *Elaeocarpus sp.,* and a *Peripentadenia sp.*

Negative tests were obtained for the following: *Aristotelia fruticosa, Elaeocarpus bifidus, E. chinensis, E. japonicus, E. sphaericus, E. syl-*

*vestris, E. ulianus, Sericola sp., Sloanea brevipes, S. dasycarpa,
S. dentata, S. grandiflora, S. guianensis, S. lasiocoma, S. purdiaea,
S. schumanni, Sloanea spp.* (3), *Vallea stipularis.*

ELATINACEAE
2 genera; 32 species

The two genera in this small family are aquatic or swamp dwellers of temperate and particularly tropic zones. They have no known economic importance.

Alkaloids are not known in the family. Three samples including two species of *Bergia, B. decumbens* and *B. glutinosa,* were tested with negative result.

EMPETRACEAE
3 genera; 5 species

These evergreens are found in north temperate mountainous regions and extend from the Arctic to the Antarctic. Some are cultivated as ornamentals.

Alkaloids are not known nor were positive tests obtained on two samples of *Ceratiola ericoides.*

EPACRIDACEAE
31 genera; 400 species

The family occurs from Indo-Malaysia to Australia with a few species found in South America. A single positive alkaloid test has been reported for a species of *Leucopogon.*

Thirty-seven samples representing 30 species in the family gave but one positive result, *Styphelia* (= *Cyathodes*) *juniperina.* Species including the following were negative: *Brachyloma ciliata, B. scortechnii, Cyathodes acerosa, C. fasciculata, Dracophyllum filifolium, D. recurvum, Dracophyllum spp.* (4), *Epacris alpina, Leucopogon albicans, L. longistylis, L. parviflorus, L. richei, Leucopogon spp.* (7), *Penta-*

chondria pumila, Richea gunnii, Sprengelia incarnata, Styphelia sua-veolens, S. tameiameiae, Trochocarpa dekockii, T. papuana.

EPHEDEACEAE
1 genus; 40 species

Once considered part of the Gnetaceae, *Ephedra* is now placed in a family of its own. A northern hemisphere genus with some representation in southern South America, *Ephedra* is the source of the ancient Chinese drug *Ma-huang* from which the alkaloid ephedrine and its relatives used in modern medicine are obtained.

One of two samples of *E. pedunculata* gave a positive test for alkaloids; samples of *E. aspera, E. trifurca,* and *E. californica* did not.

EQUISITACEAE
1 genus; 29 species

The family is cosmopolitan except for Australia and New Zealand. The majority of the species are tropical to subtropical. *Equisetum* is known as the scouring rush for its former use as a polishing agent due to the accumulation of silica in the tissues.

Alkaloids are known in the family, but in the present study, only two positive tests were obtained with 12 samples including eight species: *Equisetum ramosissimum* and one unidentified. The rest were negative: *E. arvense, E. bogotense, E. laevigatum, Equisetum spp.* (2), *E. telmateia.*

ERICACEAE
103 genera; 3,350 species

The family is cosmopolitan with the exception of deserts. It contains many of our familiar ornamentals (rhododendron, azalea, etc.), fruits (blueberries, cranberries, and relatives), briar, and wintergreen.

A few genera have been reported to give positive alkaloid tests and some of the isolated compounds have been studied. Considering the size of the family, few alkaloids have been described.

Two hundred and sixty-one samples were tested, representing 203 species, with the following positive results: *Arctostaphylos angustifolia, Bejaria racemosa, Cavendishia sp.* (1/2), *Erica baccans, Gaylussacia brasiliensis, Leucothoe editorum* (1/2), *Oxydendrum arboreum, Rhododendron homophorum, Rhododendron sp.* (1/3).

Negative tests were obtained with samples of the following: *Agapetes costata, A. sclerophylla, Arbutus glandulosa, A. laurina, A. menziesii, A. xalapensis, Arctostaphylos arguta, A. columbiana, A. conzatti, A. drupacea, A. glauca, A. lucida, A. patula, A. polyfolia, A. pungens, A. rudis, A. tomentosa, A. uva-ursi, Azalea sp., Calluna vulgaris, Cavendishia sp., Chamaedaphne calyculata, Dimorphanthera denticulifera, D. kempteriana, D. splendens, Diplycosia morobeensis, Epigaea repens, Eremia totta, Erica bauera, E. brownleeae, E. caffra, E. coccinea, E. corifolia, E. culica* var. *coronifera, E. culumiflora, E. curviflora, E. descipiens, E. densiflora, E. drakensbergensis, E. fastigata, E. gracilis, E. imbricata, E. inflata, E. johnstoniana, E. lanata, E. lucida, E. lusitanica, E. nabea, E. peziza, E. patersonia, E. plukenetii, E. regia, E. sparmannii, E. speciosa, Erica sp., E. tenella, E. versicolor, E. viridifolia, E. viridipurpurea, E. woodii, Gaultheria acuminata, G. angustifolia, G. antipoda, G. cumingiana, G. depressa, G. glaucifolia, G. hidalgensis, G. borneesis, G. hirtiflora, G. longipes, G. mundula, G. nitida, G. odorata, G. procumbens, G. pullei, G. shallon, Gaultheria spp.* (2), *Gaylussacia, G. brachycera, G. pallida, G. frondosa, G. pseudogaultheria, Gaylussacia spp.* (5), *Griesbachia rigida, Hugeria lasiostemon, Kalmia angustifolia, K. latifolia, Ledum glandulosum, Leiophyllum buxifolium, Leucothoe axillaris, L. chlorantha, L. niederleinii, L. pulchella, L. racemosa, Leucothe sp., Loiseleuria procumbens, Lyonia ferruginea, L. lingustrina, L. lucida, L. ovalifolia, Maclennia insignis, Pentapterygium serpens, Pernettya ciliaris, Philippia evansii, P. simii, Pieris taiwanensis, P. foralifolia, P. japonica, P. mariana, P. phillyreifolia, Pieris sp., Psammisia leucostoma, Rhododendron aurigeranum, R. carolinianum, R. catawbiense, R. christii, R. cinnamomeum, R. commonae, R. formosanum, R. gracilentum, R. herzogii, R. inconspicum, R. invariorum, R. konori, R. macgregoriae, R. macrophyllum, R. maddenii, R. maximum, R. micranthum, R. morii, R. nudiflorum, R. occidentale, R. oldhamii, R. ovatum, R. pachycarpon, R. phoniceum, R. pseudochrysanthum, R. rarum, R. rubropilosum, R. rubropunctatum, R. schlippenbachii,*

R. serrulatum, R. setosum, Rhododendron spp. (2), *R. tanakai, R. vaccinoides, R. viscosum, R. yelliottii, Satyria warsewicia, Sphyrospermum majus, Sympieza articulata, Vaccinium albicans, V. ambylandrum, V. amplifolium, V. arboreum, V. bracteatum, V. confertum, V. corymbosum, V. donianum, V. emarginatum, V. finis-terrae, V. geminiflorum, V. hirtum, V. ingens, V. molle, V. myrsinites, V. myrtilloides, V. ovatum, V. pallidum, V. parvifolium, V. randaiense, V. retusa, Vaccinium spp.* (5), *V. stramineum, V. striicaule, V. vacillans, Xolisma ferruginosa, X. ovalifolia, Xylococcus bicolor.*

ERIOCAULACEAE
14 genera; 1,200 species

This is essentially an American family of the tropical and warm temperate areas. A few of its members are used in decorative flower arrangements.

Alkaloids are not known in the family. In the study reported here, only one undetermined species of *Paepalanthus* gave a positive test. The following were negative: *Eriocaulon africanum, E. dectangulare, E. formosanum, E. montanum, E. novoguineense, E. ligulatum, E. papuanum, E. piorensis, E. saccatum, Eriocaulon sp., E. wightianum, Paepalanthus albo-vaginatus, P. fasciculatus, P. moldenkianus, P. planifolius, P. polyanthus, Tonina fluviatilis.*

ERYTHROXYLACEAE
4 genera; 260 species

The family is primarily tropical American and one genus (*Erythroxylum*) is known for its production of cocaine. Related alkaloids are also found; constituents have been reviewed.

Twenty-four samples, each of 23 representing a species of *Erythroxylum*, were tested to give positive results for *E. coca* (well known) as well as for *E. ellipticum, E. marginatum,* and an unidentified species. Others were negative: *E. australe, E. argentinum, E. cumanense, E. deciduum, E. floribundum, E. micranthum, E. microphyllum,*

E. ovatum, E. pictum, E. suberosum, Erythroxylum spp. (9), and *Nectaropetalum zuluense.*

EUCOMMIACEAE
1 genus; 1 species

The one species of this family is Chinese and is cultivated as an ornamental. Alkaloids have not been reported but one of three samples of *Eucommia ulmoides* gave a positive test.

EUPHORBIACEAE
321 genera; 7,950 species

This large family is cosmopolitan except for arctic areas, with centers of distribution in tropical America and Africa. It is of major economic importance as the source of rubber, tung and castor oils, a basic food crop (manihot, native to South America and introduced into Africa and southeast Asia), and familiar ornamentals (croton, poinsettia).

Several types of alkaloids are found throughout the family; their chemical and biological properties have been reviewed (Hirata, 1975). In this survey, 808 samples representing 623 species were tested. Some of these included well-known alkaloidal plants: *Acalypha indica, Alchornea cordifolia, Astrocasia phyllanthoides, Croton arnhemicus, C. centidifolius, C. draco, C. linearis, C. rhamifolius, Euphorbia atota* (1/2), *Fluegga virosa, Hymenocardia acida* (1/2), *Ricinus communis* (3/13), *Securinega virosa.*

Other species of many of the same genera likewise gave positive results: *Acalypha arvensis* (1/3), *A. ciliata, A. decumbens, Alchornea cordata* (1/2), *A. hirtella, A. laxiflora, A. rugosa, A. trimera, Alchorneopsis trimera, Andrachne decaisnei, Antidesma polyanthum, A. venenosum* (1/2) (the genus is placed in the Stilaginaceae by some taxonomists), *Bernardia interrupta* (1/3), *Bridelia mollis* (1/2), *Chrozophora sp., Clutia abyssinica, C. affinis, C. pulchella* (1/2), *Clutia sp.* (1/3), *C. swynnertonii, Croton cajucara* (2/2), *C. californicus* (1/2), *C. ciliatoglandulosis* (2/3), *C. cortesianus*

(1/5), *C. dioicus, C. fergusonii, C. flavens, C. glabellus, C. glandulosus* (1/2), *C. gossypifolia, C. gratissimus* (2/3), *C. guatemalensis, C. landleyi, C. megalobotrys,* (2/2), *C. palanostigma, C. punctatus, C. reitzii, C. rivularis, C. soliman* (1/2), *Croton spp.* (4/20), *C. steenkampiana, C. subgratissimus, Dalechampia galpinii, D. calycinum* (2/2), *D. gracile* (2/2), *D. membranaceum* (3/5), *Elaeophorbia drupifera, Endospermum chinese, Eremocarpus setigerus, Erythrococca berberidea, Euphorbia angularis, E. cyparissias* (1/3), *E. floridana* (1/2), *E. montieri, E. rectirama, Euphorbia spp.* (2/12), *E. striata, Exocecaria dallachyana, Flueggea macrocarpa, F. microcarpa, Glochidion sp., Hevea guianensis, Hyeronima laxiflora, Jatropha campestris, J. schlechteri, Lingelsheimia* (= *Drypetes*) *gilgiana, Mabea sp.* (1/2), *Macaranga barteri, Mallotus apelta, M. nepalensis, M. paniculata, Micrococca mercurialis* (1/2), *Pera anisotricha, Phyllanthus fluitans, P. orbiculatus, Phyllanthus spp.* (2/16), *Pycnocoma cornuta, Sapium jamaicensis, S. sebiferum* (1/3), *Sebastiania schottiana, Securinega ramiflora, Suregada africana, Synadenium cameronii.*

Alkaloids were not detected in the following samples: *Acalypha allenii, A. angustata, A. australis, A. brevicaulis, A. caperonioides, A. caturus, A. crenata, A. flagellata, A. fruticosa, A. glabrata, A. gracilens, A. gracilis, A. hederacea, A. langiana, Acalypha sp. aff. langiana, A. macrostachyoides, A. maerostachya, A. neptunica, A. oligodontha, A. oreopila, A. ornata, A. ostryaefolia, A. phleoides, A. psilostachys, A. rhomboidea, A. senensis, Acalypha spp.* (6), *A. stachyura, A. unibracteata, A. wilkesiana, Adelia barbinervis, Adenocline mercurialis, Adriana klotzschii, Alchornea castanaefolia, A. keelungensis, A. triplinervia, A. sidifolia, Aleurites fordii, A. moluccana, Aleurites sp., Amanoa guianensis, Amperea xiphoclada, Andrachne ovalis, Androstachys johnsonii, Antidesma ghaesembilla, A. japonica, A. parviflora, A. platyphyllum, A. pulvinatum, Antidesma sp., Aporusa chinensis, Baloghia lucida, Bernardia aspera, B. iborata, B. mexicana, B. myricifolia, Bernardia sp., Beyera leschenaultia, Bischoffia javanica, B. trifoliata, Breynia cernua, B. fruticosa, B. nivosa, B. oblongifolia, Breynia sp., Bridelia cathartica, B. duvigneaudii, B. micrantha, B. minutiflora, B. monoica, B. stipularis, Caperonia buettneriacea, Caperonia sp., Cephalocroton pueschelii, Claoxylon angustifolium, C. discolor, Cleistanthus apodus,*

C. schlechteri, C. aconitifolius, Clutia alaternoides, C. daphnoides, C. inyangensis, C. monticola, C. natalensis, C. robusta, C. rubricaulis, C. benguelensis, Cnidosculus sp., C. ureus, C. aconitifolius, Codiaeum variegatum, Coelenodendron mexicanum, Colliquaja brasiliensis, Croton antisyphiliticus, C. argyranthemus, C. brasiliensis, C. capitatus, C. chaetophorus, C. corymbalosus, C. crassifolius, C. fragilis, C. glabellus, C. heterodoxus, C. humilis, C. insularis, C. jatropha, C. lobatus, C. lundianus C. maritimus, C. matourensis, C. migrans, C. moluccanum, C. morifolius, C. myrianthus, C. niveus, C. populifolius, C. pycnocephalus, C. reflexifolius, C. sellowii, C. suberosus, C. sylvaticus, C. tiglium, C. tomentosus, C. trinitatis, C. wigginsii, Crotonopsis linearis, Ctenomeria capensis, Dalechampia micromeris, D. pentaphylla, D. roezliana, Dalechampia spp. (2), D. scandens, D. tiliaefolia, C. papuanum, Daphniphyllum sp., Dissilaria sp. aff. muelleri, Ditaxis heterantha, D. lanceolata, D. pringlei, Drypetes arguta, D. gerrardii, D. mossambicensis, Elaeophorbia abutaefolia, Endospermum macrophyllum, E. medullosum, E. moluccanum, E. myrmecophyllum, Erythrococca natalensis, Erythrococca sp., Euphorbia antisyphilitica, E. antiquorum, E. ariensis, E. avasmontana, E. brasiliensis, E. burmanii, E. caecorum, E. calyculata, E. celastroides, E. cinerascens, E. ciparicea, E. clusiaefolia, E. colletioides, E. commutata, E. cooperi, E. crotonoides, E. cyparissioides, E. cythophora, E. dentata, E. dioscorioides, E. drummondii, E. elegans, E. epicyparissus, E. erythrina, E. evansii, E. eylesii, E. fulva, E. grandicornis, E. graminea, E. heterophylla, E. hillebrandii, E. hirta, E. hyssopifolia, E. inaequilatera, E. ingens, E. kraussiana, E. leucocephala, E. lingulata, E. lupatensis, E. maculata, E. macgillivrayi, E. matabelensis, E. mauritanica, E. melandenia, E. mesembryanthemifolia, E. nerifolia, E. nesmanii, E. peplus, E. phylloclada, E. pilulifera, E. polycarpa, E. prunifolia, E. pseudocactus, E. pubiglans, E. pulcherrima, E. pulvinata, E. rhombifolia, E. rudis, E. schinzii, E. schlechtendalii, E. serrula, E. serrulata, E. stolonifera, E. tettensis, E. thymifolia, E. tirucalli, E. tomentulosa, E. torrida, E. transvaalensis, E. triangularis, E. tuberosa, E. villifera, Excoecaria allagocha, E. bicolor, E. orientalis, Garcia nutans, Gelonium aequoreum, Glochidon ereiocarpum, Gymnanthes lucida, Hevea nitida, Homalanthus populifolius, Homalanthus sp., Hura crepitans, Hyaenache globosa, Hyeronima caribea, Hyeronima sp., Hymeno-

cardia ulmoides, Jatropha aff. *curcas, J. curcas, J. dioica, J. erythro-poda, J. gossypii, J. gossypifolia, J. liebmannii, J. multifida, J. multi-loba, Jatropha spp.* (3), *J. variifolia, J. zeyheri, Joannesia princeps, Julocroton sp.*, *Kirganelia reticulata, Mabea angustifolia, Mabea* aff. *occidentalis, M. nitida, M. occidentalis, M. subsessilis, Macaranga capensis, M. quadriglandulosa, Macaranga spp.* (5), *Mallotus japo-nicus, M. oppositifolius, M. philippensis, M. polyadenus, M. repan-dus, M. ricinoides, Mallotus sp.*, *Manihot angustiloba, M. caudata, M. colimensis, M. dulcis, M. esculenta, M. gracilis, M. gracipes, M. grahamii, M. pringlei, Manihot spp.* (6), *M. utilissima, Maprou-nea brasiliensis, M. guianensis, Margaritaria nobilis, Melanolepsis multiglandulosa, Micrandra sp.*, *M. syphonioides, Micrococca ca-pensis, Oldfieldia dactylophylla, Omphalea* cf. *queenslandicus, Pachystroma longifolium, Pausandra morisiana, Pedilanthus calca-satus, P. macradenius, P. spectabilis, P. tithymaloides, Pera districho-phylla, P. obovata, P. schomburgkiana, Pera sp.* (2), *Petalostigma banksii, P. quadriloculare, Phyllanthus acuminatus, P. brasiliensis, P. burchelli, P. cochinchinensis, P. compressus, P. emblica, P. engleri, P. flexuosus, P. graminicola, P. juglandifolius, P. kirkianus, P. mader-aspatensis, P. micrandrus, P. meyerianus, P. muelleranus, P. myrtifo-lius, P. nuriri, P. pendulus, P. pentandrus, P. piscatorum, P. polygo-noides, P. pulcher, P. reticulatus, P. urinaria, P. welwitschianus, Picrodendron baccatum, Pimeleodendron amboinicum, P. papua-num, Piranhea trifoliata, Pogonophora schomburgkiana, Poinsettia* (= *Euphorbia*) *cyanthopora, P. pulcherrima, Polyandra sp.*, *Pseudo-lachnostylis maprouneifolia, Putranjiva roxburghii, Richeria austra-lis, R. grandis, Ricinocarpus pinifolius, Sapium acuparium, S. cornu-tum, S. glandulatum, S. hippomane, S. integerrinum, S. japonicum, S. macrocarpum, S. sellowianum, Sapium spp.* (3), *Sauropus rostra-tus, Sebastiania corniculata, S. fruticosa, S. hispida, S. klotzschiana, S. pavoniana, S. schottiana, Sebastiania sp.*, *Securinega suffruticosa, Senefeldera ducke, Siphonia globulifera, Spathiostemon javense, Sphyranthera odorifera, Spirostachys africana, Stillingia aquatica, S. sanguinolenta, Stillingia spp.* (4), *S. sylvatica, S. zelayensis, Sur-egada sp.*, *Synadenium cupulare, Tetracoccus dioicus, Tragia gard-neri, T. nepetaefolia, Tragia spp.* (3), *T. volubilis, Trewia nudiflora, Uapaca benguelensis, U. kirkiana, U. nitida, U. robynsii.*

REFERENCE

Hirata, Y., *Pure and Applied Chemistry 41* (1975) p. 175.

EUPOMATIACEAE
1 genus; 2 species

This is a small family of Australia and New Guinea of no known economic importance.

Aporphine alkaloids have been identified in *Eupomatia laurina*, which also gave a positive test in this study.

F

FAGACEAE
7 genera; 1,050 species

The beeches are found mostly in the northern hemisphere from temperate to subtropical habitats. One genus is pantropical; another approaches the Antarctic. Economically important species include varieties of lumber, edible chestnuts, and some ornamentals.

There is one report of alkaloids, but this may have been due to the presence of amino acids otherwise known in the family. Four positive tests were obtained in the present study: *Quercus centralis* (1/2), *Q. laurina* (1/3), *Q. microphylla* (1/5), and *Q. prinus* (1/5).

Negative tests were obtained with the following: *Castanea acuminatissima, C. ashei, C. crenata, C. dentata, C. pumila, Castanopsis borneensis, C. carlesii, C. hystrix, C. kawakamii, C. kusanoi, C. sempervirens, Cyclobalanopsis paucidenta, C. glauca, C. morii, C. seikooruotan, Fagus americanus, F. grandifolia, Lithocarpus amygdalifolius, L. calicoris, L. monopetala, L. novoleontis, L. rotundifolia, L. schaffneri, L. vitiana, L. densiflora, Lithocarpus spp.* (2), *Nothofagus cliffortioides, N. grandis, N. truncata, Pasania* (= *Lithocarpus*) *brevicaudata, Quercus acutissima, Q. affinis, Q. agrifolia, Q. alba,*

Q. borealis, Q. castnea, Q. chialmensis, Q. chihuahuensis, Q. chry-solepsis, Q. chrysophylla, Q. coccinea, Q. conspersa, Q. couspeosa, Q. crassifolia, Q. crassipes, Q. discolor, Q. diversifolia, Q. dougla-sii, Q. dumosa, Q. durata, Q. eduardi, Q. fulva, Q. furfuracea, Q. gambelli, Q. garryana, Q. glabrata, Q. glaucum, Q. hartwegii, Q. heterophylla, Q. ilex, Q. ilicifolia, Q. imbricaria, Q. incana, Q. kelloggii, Q. laevis, Q. lanigera, Q. lobata, Q. macrocarpa, Q. macrophylla, Q. marilandica, Q. mexicana, Q. michauxii, Q. myrsinaefolia, Q. oblongifolia, Q. oleoides, Q. palustris, Q. per-seaefolia, Q. phellos, Q. polymorpha, Q. potosina, Q. pringlei, Q. prinoides, Q. repanda, Q. robur, Q. rubra, Q. rugosa, Q. rugulo-sa, Q. saulii, Q. serrata, Q. shumardii, Quercus sp., Q. stellata, Q. tinkhami, Q. transmontana, Q. turbinella, Q. variablis, Q. veluti-na, Q. wislizeni, Q. xalapensis, Shiia (= Castanopsis) cuspidata, S. longicaudata, S. sieboldii.

FILICOPSIDA (FERNS)

The classification of ferns appears to be complex. Whether they should be lumped into one or two large groups or split into as many as 36 families seems not yet a matter of agreement among taxonomists. In view of the relatively few samples tested in each of the several "families," and of the fact that very few indications of the presence of alkaloids in the ferns have been noted, they are treated here as a group, using family designations as given by the suppliers of the test samples and noting changes as given by Mabberley. Genera are listed with the species tested and all tests were negative unless otherwise noted.

Adiantaceae
56 genera; 1,150 species; cosmopolitan

Adiantum alenticum, A. bellum, A. capillus-veneris, A. cauda-tum, A. coccinum, A. cuneatum, A. flabellatum, A. pedatum, A. formosanum, Adiantum spp. (2), *Pityrogramma calomelanos. Adiantum philippense* was positive (1/2).

The family includes Pteridaceae and Vittariaceae inter alia according to some authorities.

Aspidiaceae
Sometimes included in Aspleniaceae*

Angiopteris (= *Onoclea*) *lygoliifolia, A. suboppositifolia, Dryopteris amplissima, D. decursivo-pinnata, D. erythrosa, D. goldiana, D. gymnosora, D. intermedia, D. ludoviciana, D. marginalis, D. noveboracensis, D. oligophebia-lasioca, D. paleacea, D. patens, D. parasitica, D. schimperiana, D. scottii, D. thelypteris, Onoclea sensibilis* (2/2), *Paranema cycatheoides, Polybotria cervina, Polystichum aculeatum, P. amabilis, P. falcatipinnum, P. hancockii, P. lepidocaulon, P. montevidense, P. munitum, P. nepalense, P. vestitum, Pteris aqualinia, P. cretica, P. dispar, P. ensiformis, P. fauriei, P. semipinnata, P. semipinnata dispar, P. vittata, Tectaria heracleifolia, T. sugtriphylla, Woodsia obtusa.*

An unidentified species of *Dryopteris* and one of *Pteris* were positive.

Aspleniaceae
78 genera; 2,200 species

These include several other families by some authorities: cosmopolitan epiphytes or rock plants.

Asplenium bulbiferum, A. ensiforme, A. flaccidum, A. laserpitiifolium, A. lucidum, A. prolongatum, Asplenium sp., A. wightii, Ceterach officinarum, Ctenitis apiciflora, C. dawadamii, C. eatonii, C. kawakamii, C. subglandulosa, C. trichorachis.

Trichomanes (= *Asplenium*) *makinoi*, in Hymenophyllaceae by some authorities, was positive.

Athyriaceae*

Included in the Aspleniaceae. *Athyrium angustum, A. australe, A. arisanense, A. lanceum, A. oppositipinnum, Diplazium lanceum, D. dilatatum, D. bantamense, D. kawakamii, D. maximum, D. phalelepis, D. pseudoederleinii.*

---

*When family status is in doubt, the number of genera and species are also in question; therefore, the numbers are not given here. This note applies where asterisk appears.

Blechnaceae
10 genera; 260 species

Sometimes tree ferns or climbers. *Blechnum capense, B. discolor, B. fluviatile, B. glandulosum, B. imperiale, B. meridense, B. nipponicum, B. orientale, B. serratulum, Blechnum sp.*, *Sadleria cyatheoides, Stenochlaena palustris, Woodwardia areolata, W. orientalis, W. unigemmata, W. virginica.*

Cyatheaceae
2 genera; 625 species

Tree ferns of warm to tropical regions, often in montane forests. *Alsophila corcovadensis, A. glabra, Cibotium barometz* (now in Thyrsopteridaceae), *Cyathea dealbata, C. dregei, C. medullaris, C. taiwanensis, Dicksonia barometz, D. squarrosa* (both in Dicksoniaceae), *Gymnosphaera formosana, G. podophylla, Hemitelia (= Cyathea) capensis.*

Davalliaceae
13 genera; 220 species

Found in warm and tropical regions; mostly epiphytic. Sometimes included in Oleandraceae. *Arthropteris obliterata, Davallia mariosii, Humata parvula, H. repens, Hymolepis punctata, Leucostegia immersa.*

Dennstaedtiaceae
24 genera; 410 species; cosmopolitan

Includes Lindsaeaceae and Monacosoraceae by some authorities. *Dennstaedtia hirsuta, D. punctilobula, D. scabra, D. scandens, Histiopteris incisa, Hypolepis puncata, Microlepis hookeriana, M. setosa, Paesia scuberula, Pteridium aquilinum.*

Dicksoniaceae
2 genera; 26 species

In tropical America, the southwest Pacific, and the island of St. Helena. Often included in Cyatheaceae. *Dicksonia barometz, D. squarrosa.*

Gleicheniaceae
4 genera; 410 species

Of tropical and warm south temperate zones. *Dicranopteris linearis, D. splendida, Gleichenia bolanica, G. cunninghamii, G. dicarpa, G. dichotoma, G. erecta, G. linearis, Gleichenia sp., G. venosa, G. vulcania, G. warburgi, Micropteris (= Dicranopteris) glauca, M. longissima.*

Graimitidiaceae
11 genera; 500 species

Of cloud forests and tropical and Australian mountains. *Ctenopteris curtisii, C. obliquatus.*

Gymnogrammaceae*

Included in Adiantaceae by some authorities. *Coniogramme intermedia, Hemionitis elegans.*

Hymenophyllaceae
33 genera; 460 species

Filmy ferns of tropical and some temperate regions. Some taxonomists include these with the Aspleniaceae. *Trichomanes makinoi* gave a positive alkaloid test; the rest were negative: *Crepidomanes makinoi, Mecodium polyanthos, Selenodesmium obscurum, Vandenboschia radicans.*

Isoetaceae
2 genera; 77 species

Cosmopolitan aquatics except in the islands of the Pacific; related to *Lycopodium* and *Selaginella*. *Isoetes novoguineensis*.

Lindsaeaceae*

In Dennstaedtiaceae by some authorities. *Lindsaya barbiculata, L. cultrata, L. orbiculata, Odontosoria chinensis, Sphenomeris chinensis, S. chusana, Stenoloma* (= *Sphenomeris*) *chrisanum*.

Lomariopsidaceae*

In Aspleniaceae by some authorities. *Elaphoglossum spp.* (2), *E. wagneri*.

Monacosoraceae*

Usually included in Dennstaedtiaceae. *Monachosorum subdigitatum*.

Marattiaceae
7 genera; 100 species

Of tropical and warm zones. *Angiopteris* (= *Onoclea*) spp. (2), included in Aspidiaceae.

Marsiliaceae
3 genera; 70 species

Warm and tropical areas. *Marsilia brownii*.

Oleandraceae*

Included by some in the Davalliaceae. *Oleandra wallichii, Arthropteris obliterata, Nephrolepis auriculata, N. biserrata, N. exaltata, N. hirsutula.*

Ophioglossaceae
4 genera; 65 species

Mainly temperate herbs with some tropical epiphytes. *Botrichium virginianum, Ophioglossum vulgatum.*

Osmundaceae
3 genera; 19 species

Tropical and temperate ferns often cultivated as ornamentals. *Liptopteris superba, Osmunda cinnamomea, O. datonianum, O. japonica, O. regalis.*

Parkeriaceae
1 genus; 4 species

Floating ferns of warm and tropical regions; some are eaten. *Ceratopteris thalictroides.*

Plagiogyriaceae
1 genus; 37 species

Found in eastern Asian and American forests on mountain ridges. *Plagiogyria falcata, P. formosana.*

Polypodiaceae
52 genera; 550 species

A large group of ferns that, at one time or another, included many of the families later separated from this cosmopolitan, but primarily tropical, family. It supplies some foods, medicines, and cultivated ornamentals.

Arthromeris lehmannii, Cheilanthes angusifolia, Cyclophorus lingua, C. alatellus, C. linearifolius, Drynaria rigidula, Lemnaphyllum subrostratum, Lepisorus heterolepis, L. kawakamii, L. infraplanicostalus, L. monilisorus, L. obscurivenulosus, L. thunbergianus, Loxogramme ramotifrondigera, L. salicifolia, Onychium japonicum, O. soliculosum, Pessopteris crassifolia, Phymatodes scolopendria, P. diversifolium, Pleuropeltis sp., Polypodium achrosticoides, P. diversum, P. formosanum, P. fortunei, P. juglandifolium, P. polypodoides, P. resei, Polypodium spp. (4), *P. taiwanianum, P. tectum, P. virginianum, Pseudodrynaria coronans, Pyrrosia adnascens, P. mollis, P. polydactylus, P. sheareri.*

Psilotaceae
2 genera; 5-9 species

Tropical and subtropical epiphytes or rock plants. *Psilotum nudum.*

*Pteridaceae**

Included in Adiantaceae. *Acrostichum aureum.*

Schizaceae
4 genera; 150 species; subcosmopolitan

Mostly in warm to tropical areas. *Anemia spp.* (5), *Lygodium japonicum, L. microphyllum, L. reticulatum, Mohria caffrorum, Schizaea dichotoma, S. digitata, S. malaccana.*

*Sinopteridaceae**

In Adiantaceae by some authorities. *Aleuritopteris farinosa, Pellaea falcata, P. nitidula, Cheilanthes tenuifolia.*

Thelypteridaceae
30 genera; 900 species; subcosmopolitan

Abacopteris triphylla, Thelypteris decursivo-pinnata, T. penn-igera.

Thrysopteridaceae
3 genera; 20 species

Of Macronesia and the tropics generally except Africa. Some include *Dicksoniaceae* (q.v.). *Cibotium barometz.*

Vittariaceae*

Included in Adiantaceae. *Vittaria flexuosa, Vittaria sp., V. taenio-phylla.*

FLACOURTIACEAE

This is a family of pantropical distribution, some members of which are used as ornamentals and a few others for their edible fruit. Perhaps best known is *Hydnocarpus* for chalmoogra oil which, for many years, was the only treatment for leprosy.

The occurrence of alkaloids in the family is sporadic; fairly well known are those of *Ryania* for their insecticidal activity. Positive results were obtained with 16 species, including *Casearia sylvestris,* known to contain alkaloids: *Casearia grandiflora* (1/2), *C. lasiophyl-la, C. nitida, Casearia spp.* (3/9), *Dovyalis caffra* (2/2), *Dovyalis sp., Homalium foetidum, Neopringlea integrifolia, Ptychocarpus* (= *Neoptychocarpus*) *opodanthus* (1/3), *Ryania angustifolia, Scopolia zeyheri, Trimeria trinervus, Xylosma ellipticum, Zuellania guidonia.*

The following were negative: *Abatia mexicana, A. tomentosa, Aberia sp., Aphloia theiformis, Asteriastigma* (= *Hydnocarpus*) *macrocarpa, Baileyoxylon lanceolatum, Banara guianensis, B. to-mentosa, Caloncoba suffruticosa, Carpotroche crispidentata, C. lon-*

gifolia, Casearia clutiaefolia, C. decandra, C. dolichophylla, C. erythrocarpa, C. esculenta, C. guianensis, C. inaequilatera, C. juvitensis, C. junodii, C. nigricans, C. pringlei, C. rhynochophylla, Flacourtia cataphracta, F. indica, F. ramontchi, F. rukam, F. sepiaria, F. zippelii, Hasseltia mexicana, Homalium dentatum, H. pedicillatum, H. trichostemon, H. zeylanicum, Hydnocarpus anthelmintica, Idesia polycarpa, Kiggelaria africana, Laetia procera, L. suaveolens, Lindackeria latifolia, Lindackeria sp. aff. *maynesis, Lightfootia abyssinica, L. albens, L. huttoni, L. parvifolia, L. peratifolia, L. tenella, Mayna toxica, Muntingia calabura, Oncoba laurina, O. spinosa, Pangium edule, Paropsia brazzeana, Prokia crucis, Rawsonia lucida, Ryania pyrifera, R. speciosa, Ryparosa calotricha, Ryparosa sp.* cf. *javanica, Trichadenia philippinensis, Trimeria grandiflora, Xylosma celastrium, X. ciliatifolium, X. flexuosum, X. glaberrimum, X. palmeri, Xylotheca kraussiana.*

FLAGELLARIACEAE
2 genera; 4 species

This is a small family of the Old World tropics of no particular economic importance. The genus (*Flagellaria*) had been reported to contain alkaloids but four species were found negative in this study: *Flagellaria guianensis, F. indica, Flagellaria sp.,* and *Joinvillea elegans.* This last species is now considered in a family of its own by some authorities.

FOUQUIERIACEAE
1 genus; 11 species

Fouquieria is a Mexican genus extending northward into the southwestern portions of the United States. A few species have limited use as ornamentals.

Alkaloids are not known nor were positive tests obtained on six samples representing *Fouquieria formosa, F. shrevei,* and *F. splendens.*

FRANKENIACEAE
3 genera; 30 species

The family occurs worldwide but is represented primarily in the Mediterranean area. A limited number are cultivated as novelties. *Frankenia* was reported to contain alkaloids but one sample of each of two unidentified species failed to give a positive test.

FUMARIACEAE
18 genera; 450 species

Long considered a subfamily of the Papaveraceae, this family is of Old World distribution, chiefly of temperate Asia, with four genera in South Africa and three in the United States. A few ornamentals have economic importance.

Alkaloids are found in several genera of the family; they are chemically similar, and in many cases identical, to those found in the Papaveraceae. One unidentified species of *Corydalis* was negative although the genus is one of the alkaloidal genera of the family.

G

GARRYACEAE
1 genus; 13 species

This is a family of western North and Central America with one species in Jamaica. A few are used as ornamentals.

Alkaloids are known in the family. In the present study, two species of *Garrya*, *G. laurifolia* and *G. veatchii*, were positive, the latter known to be alkaloidal. Two others were negative: *G. longifolia* and *G. ovata*.

GENTIANACEAE
74 genera; 1,200 species

Gentians have worldwide distribution with concentration in the temperate zones. Many members of the family are used as ornamentals.

Alkaloids are known but from very few genera. In this study, the following were positive: *Chenolanthus alatus, Chironia baccifera, C. melampyrifolia, C. tetragonia, C. transvaalensis, Exochaenium* (= *Sabea*) *grande, Frasera neglecta, Gentiana acaulis, G. adsurgens* (2/3), *G. andrewsii, G. bisetae, G. lutea, G. purdomii, G. spathacea, G. superba, G. verna, Halenia brevicornis* (1/4), *Orphium frutescens, Sabatia difformis.*

Negative tests were obtained on the following species: *Calolisianthus pedunculatus, C. speciosus, Calolisianthus sp., Chelonanthus alatus, Chironia palustris, Coutoubea spicata, Curtia conferta, Erythraea* aff. *chirinoides, E. tetraniera, Exacum perrottetii, E. tetragonum, Gentiana billidifolia, G. cinereifolia, G. cruciata, G. decumbens, G. diemensis, G. formosana, G. juniperina, G. parvifolia, G. porphyrio, G. mexicana, G. septemfida, G. walujewi, Gentianella amarella, Halenia hintoni, H. plantaginea, Halenia spp.* (2), *Limnanthemum christatum, L. humboldtianum, Lisianthus brittonii, Schultesia guianensis, Swertia randaiensis.*

GERANIACEAE
14 genera; 730 species

The Geraniaceae have wide distribution in both the temperate and tropical regions of both hemispheres. The family is known for its ornamentals and fragrant oils.

A few alkaloid-positive species have been recorded, including *Geranium sanguineum,* which was found positive here, but they appear to be uncommon considering the size of the family. *Geranium potentillaefolium* (1/2), *Pelargonium burtoniae* (2/2), *P. inquinans, P. lateripes, P. scabrum,* and *P. zonale* (2/2) were likewise positive.

The following plants were alkaloid-negative: *Erodium circutarium, E. crinitum, Geranium maculatum, G. aristisepalum, G. bellum, G. bicknellii, G. carolinianum, G. hayatanum, G. hernandesii, G. incanum, G. kerberi, G. liliacium, G. mexicanum, G. nyassense, G. ocellatum, G. ornithopodum, G. potentilloides, G. purpurascens, G. robertianum, G. schiedeanum, G. seemanni, Geranium spp.* (3), *G. vulcanicola, Monsonia biflora, M. burkeana, M. ovata, M. speciosa, Monsonia sp., M. umbellata, Pelargonium alchemilloides,*

P. alterans, P. angulosum, P. aconitiphyllum, P. bechuanicum, P. capitatum, P. candicans, P. coranspifolium, P. cordatum, P. elegans, P. flavum, P. graveolens, P. grossularioides, P. hirtum, P. hermanniaefolium, P. laevigatum, P. lobatum, P. moreanum, P. myrrhifolium, P. ovale, P. panduraeforme, P. peltatum, P. quercifolium, P. radula, P. rehmannii, P. reniforme, Pelargonium sp., P. saniculaefolium, P. sublingosum, P. vitifolium, Sarcocaulon patersonii, Viviania rubriflora (now in a family of its own, Vivianiaceae).

GESNERIACEAE
146 genera; 2,400 species

The family is represented primarily in the tropical and subtropical areas of both hemispheres and is important economically for its ornamentals, some especially for rock gardens.

Only a few genera have been recorded as alkaloidal. In this survey, one positive result was obtained: *Rhabdothamnus solandri.* The remainder were negative: *Achimenes grandiflora, A. heterophylla, A. pulchella, Achimenes sp.* (2), *Aeschynanthus ramosissima, Alloplectus patrisii, A. strigosus, Besleria glabra, Besleria sp., Boea swinhoi, Chirita bicornuta, B. urticaefolia, Columnea erythrophaea, C. schiediana, Columnea sp., Corytholoma sp., Cyrtandra sp., Drymonia serrulata, Hypocyrta* (= *Nematanthus*) *tessmannii, H. maculata, Isanthera discolor, Kohleria deppeana, K. fruticosa, K. hirsuta, K. hondensis, K. longifolia, K. martensii, Lysionotus warleyensis, Paliavana prasinata, Paliavana sp.* (2), *Rechsteineria curtiflora, R. spicata, Rhynchoglossum hologlossum, R. obliquum, Streptocarpus parviflorus.*

GINKGOACEAE
1 genus; 1 species

A single species, *Ginkgo biloba*, is the only one surviving since Jurassic times. Once worldwide, the genus is now confined to temple gardens in China except where introduced elsewhere.

It contains a quinoline carboxylic acid derivative and has been recorded as alkaloidal. Two of seven samples gave positive tests in this study.

GLOBULARIACEAE
10 genera; 250 species

This Old World, primarily Mediterranean, family has no known economic importance, nor is it known for the presence of alkaloids. It has been formed largely by a combination of genera formerly assigned to the Scrophulariaceae.

When tested, the following were positive: *Dischisma erinoides* (3/3), *Hebenstretia dentata, Selago compacta, S. corymbosa, S. elata, S. holubii, S. hyssopifolia* (2/2), *S. natalensis* (2/2), *S. spuria* (2/2), *S. thunbergii, S. verbenacea, Walafrida genicula, W. saxtilis, W. synnertonii.*

Negative tests were obtained with the following: *Agathelpis angustifolia, Globularia cordifolia, Hebenstretia comosa, H. dentata, H. fruticosa, Selago fruticosa, S. glutinosa, S. longipedicillata, Selago spp.* (2), *S. thomsonii, S. triquetra.*

GNETACEAE
1 genus; 28 species

The one tropical genus includes lianas and sometimes trees and shrubs. There are reports of alkaloids in *G. indicum* and *G. parvifolium* but 11 samples which included four species were negative in the tests conducted here: *G. gnemon, G. latifolium, G. nodiflorum, G. nodosum.*

GOODENIACEAE
16 genera; 430 species

This is generally considered an Australasian family, but one genus, *Scaevola,* has pantropical distribution along coastal areas of both hemispheres.

Alkaloids have been found in a few genera of the family, but these may be artifacts arising from the use of ammonia during isolation. Twenty-seven samples were tested to give the following positive results: *Goodenia rotundifolia, Scaveola gaudichaudiana*—both of which were known to be positive—and *Scaveola lensevestia, S. montana, S. plumieri, S. oppositifolia, S. sericiea, Scaveola sp.*

Negative were: *Dampiera discolor, D. purpurea, Goodenia armitiana, G. bellidifolia, G. ovata, G. ramelii, G. stelligera, Leschenaultia biloba, Scaveola albida, S. hispida, S. nitida, S. ovalifolia, S. parvifolia, Scaveola spp.* (4), *Velleia paradoxa.*

GRAMINEAE
635 genera; 9,000 species

The grasses are the most widely distributed of the plant families and of the greatest importance in furnishing the basic foods for humans and animals (rice, wheat, corn, oats, etc.) and many derived products (e.g., oils, alcohol, paper).

Many alkaloids have been characterized in this large family, yet many genera appear to await investigation for these substances. In this survey, the following gave positive tests: *Agropyron repens, Andropogon schoenanthus, Anthoxanthum odoratum, Cymbopogon narduus, C. citratus* (previously known), *Glyceria obtusa, Hordenum leporium, Ichanthus vivins* (2/2), *Lolium perenne* (2/2) (previously known), *Phalaris arundinacea* (1/2) (previously known), *Spinifex littoreus.*

The majority of the species tested were negative: *Agropyron ciliare, Agrostis alba, A. avenacea, Alopecurus aequalis, A. geniculatus, Andropogon gerardi, A. myrtiflorus, A. microstachyus, A. saccharoides, A. scorparius, Andropogon sp., A. virginicus, Apluda mutior, Aristida chinensis, Arrhenantheum elatius, Arthraxon hispidus, Arundinaria niitakayamensis, Arundinella setosa, Avena fatua, Axonopus siccus, Bambusa dolichoclada, B. oldhami, B. pervariabilis, Bambusa sp., Bothrichloa ischaemum, Bouteloua chasei, B. curtipendula, Brachiaria distachya, B. reptans, Brachypodium formosanum, Briza major, B. media, B. minor, B. rotundata, Briza sp., Bromus catharticus, B. commutatus, B. diandrus, B. inermis, B. rigidus, B. rubens, B. unioloides, Calamochloa filifolia, Capillipedium parviflorum, C. glabrum, C. kawashotense, Cenchrus calyculatus, C. echinatus, C. pauciflorus, C. tribuloides, Chloris virgata, Chusquea oligophylla, Coix distichum, C. lachryma-jobi, C. ma-yuen, Cryptococcum patens, C. tortilis, Cynodon dactylon, C. semiundulata, Danthoni mexicana, D. raoulii, Deschampsia klossii, Digitaria ascendens, D. chinensis, D. henryi,*

D. logifolia, D. longiflora, D. magna, D. microbanche, D. sanguinalis, D. scolorum, D. sericea, D. schimadana, D. violascens, Diplachne fuscata, Distichlis spicata, D. stricta, Echinochloa colonum, E. crusgalli, Ehrhartia calycina, Eleusine indica, Elymus cinereus, Elymus sp., E. villosus, Enteropogon gracilior, Eragrostis amabilis, E. bulbifera, E. cylindrica, E. diffusa, E. pilosa, E. pubulifera, Eremochloa ciliaris, E. colunum, E. opiuroides, Eriochloa procera, Eularia praemosa, E. viminea, Festuca parvigluma, F. parvilimba, Guadua angustifolia, Hackelochloa granularis, Heteropogon contortus, Hierochloe odorata, H. redolens, Holcus lanatus, H. mollis, H. brachyantherum, H. jubatum, H. vulgare, Hystrix patula, Imperata brasiliensis, I. ciliare, I. cylindrica, Isachaeum crassipes formosanum, I. globosa, I. muticum, I. setaceum, Lagurus ovatus, Lasiacus divaricata, L. latifolium, L. rascifolia, L. rugelii, L. sloanei, Leersia hexandra, Leptaspis urceolatum, Leptochloa chinensis, L. dubia, Lepturus repens, Lolium multiflorum, Lopatherum gracile, L. elatum, Melica frutescens, M. onoei, M. racemosa, M. sarmentosa, Merostachys ternata, Microstegium ciliatum, Miscanthum sinensis, Muhlenbergia repens, M. rigida, M. implicata, Nastus productus, Nothodanthonia setifolia, Olyra latifolia, O. micrantha, Oplismenus compositus, O. formosanus, O. undulatifolia, Oryza latifolia, O. perennia, O. sativa, Oryzopsis hymenoides, O. pungens, Ottochloa nodosa, Panicum aquaticum, P. clandestinum, P. distichum, P. hallii, P. incomtum, P. indicum, P. patens, P. repens, P. trichoides, P. virgatum, P. xanthophysum, P. zizanioides, Paspalum orbiculare, P. conjugatum, P. dasytrichum, P. dilatatum, P. formosanum, P. lineare, P. longifolium, P. orbiculare, P. plicatulum, P. scrobiculatum, Paspalum spp. (3), P. thunbergii, Pennisetum alopecuroides, P. latifolium, P. villosum, Perotis indica, Phalaris californica, P. caroliniana, P. tuberosa, Phragmites communis, P. karka, Phyllostachys bambusoides, P. aureus, Phyllostachys sp., Piptochaetium fimbriatum, Poa acroleuca, P. annua, P. caespilosa, P. epileuca, P. gracillima, Pogonantherum paniceum, Polypogon higecaweri, P. maritimus, Rhynchetytrum roseum, Rothboellia exaltata, Saccharum spontaneum, Sacciolepis myosuroides, Schizachyrium sanguineum, Setaria fiberii, S. geniculata, S. glauca, S. griesbachii, S. palmifolia, S. poiretiana, Setaria sp., Sitanion hanseni, Sorghastrum nutans, Sorghum bicolor,

S. halopense, S. nitidum, S. propinquum, S. sudanense, Spartina alternifolia, S. patens, Sphaerocaryon malaccense, Spinifex squarrosus, Sporobolus argutus, S. diander, S. elongatus, S. terrostris, S. indicus, S. virginicus, S. constricta, S. ichu, S. vaseyi, Thaumastochloa cochinchinensis, Thuarea involuta, Thysanolaena latifolia, Trachypogon montufori, T. secundus, Tragus berteronianus, Tricholaena rosa, Tridens pulchellus, Triodia flara, Trisetum dynexioides, Vulpia bromoides, Zea mays, Zizania aquatica.

GROSSULARIACEAE
23 genera; 340 species

This is a cosmopolitan family which, other than ornamentals, furnishes a few edible fruits (e.g., gooseberries, currants).

Alkaloids have been reported present in a few species. In this study, the following were positive: *Carpodetus serratus, Montinia caryophyllacea* (sometimes assigned to a family of its own, Montiniaceae), *Ribes americanum, R. malvaceum.*

Negative tests were obtained for the following: *Argophyllum sp., Carpodetus abroreus, C. major, Choristylis rhamnoides, Cuttsia viburnea, Escallonia grahamiana, E. montevidensis, Escallonia spp.* (3), *Itea arisanensis, I. chinensis, I. oldhamii, I. virginica* (Itea is in its own family, Iteaceae, by some authorities), *Phyllonoma laticuspis, Pterostemon mexicanus, Quintinia sieberi, Quintinia sp., Ribes affine, R. aureum, R. fasciculatum, R. indecorum, R. montigerum, R. neglectum, R. nevadense, R. odoratum, R. pringlei, R. roezlii, R. speciosum.*

GUTTIFERAE
47 genera; 1,350 species

The Guttiferae of tropical Central and South America and the West Indies are a source of ornamentals or edible fruits (mammee apple, mangosteen). The genus *Hypericum*, placed in a separate family (Hyperiaceae) by some authors, is maintained in the Guttiferae here.

Positive alkaloid tests have been recorded for about four genera in the family, though one source suggests that alkaloids are "not yet known." In the study reported here, a total of 153 samples were tested including 117 species with but four positive results: *Hypericum cistifolium, H. galioides, Moronobea coccinea, Symphonia globulifera* (1/6).

The following were negative: *Ascyrum* (= *Hypericum*) *hypericoides, A. tetrapetalum, Calophyllum australianum, C. brasiliensis, C. antillanum, C. inophyllum, Calophyllum spp.* (2), *Caraipa grandiflora, Caraipa sp., Clusia columnaris, C. ellipticifolia, C. grandiflora, C. insignis, C. mexicana, C. microstemon, C. minor, C. orthoneura, C. parvicapsula, C. polycaphala, C. rosea, C. spathulaefolia, Clusia spp.* (10), *Cratoxylon lingustrinum, Crookea* (= *Hypericum*) *microcephala, Decaphalangium peruvianum, Garcinia dulcis, G. gerrardii, G. livingstonei, G. mangostana, G. multiflora, G. pseudoguttifera, Garcinia spp.* (2), *G. tinctoria, G. warrenii, G. xanthochymus, Harungana madagascariensis, Hypericum densiflorum, H. aethiopicum, H. brasiliensis, H. connatum, H. formosanum, H. frondosum, H. gentianoides, H. habbenmense, H. japonicum, H. lalandii, H. lanceolatum, H. macgregorii, H. mutilum, H. nepalense, H. pallidum, H. papuanum, H. patulum, H. perforatum, H. prolificum, H. punctatum, H. randaiense, H. revolutum, H. roeperianum, Hypericum spp.* (7), *H. virginicum, Kielmeyera coriacea, Kielmeyera spp.* (8), *Komana* (= *Hypericum*) *patula, Mammea americana, Marila grandiflora, Oedematopus dodecandrum, Pentaphalangium sp., Platonia insignis, Psorospermum febrifugum, Rheedia acuminata, R. edulis, R. gardneriana, Symphonia globulifera, Tovomita chorpiana, T. eggersii, Tovomita sp., T. stigmotosa, Vismia angusta, V. baccifera, V. cayennensis, V. guianensis, V. japurensis, V. parviflora, Vismia spp.* (4).

H

HAEMODORACEAE
16 genera; 85 species

Representatives of the family are found in tropical North and South America, Australia, New Guinea, and South Africa. Some are cultivated for cut flowers.

The family is known for a variety of red pigments, some of which in e.g., *Lacnanthes*, are nitrogen-substituted, but alkaloids in the strict sense are seldom reported. Seven samples representing seven species gave two positive results: *Wachendorfia thyrsiflora* and *Xiphidium coeruleum.*

The remainder were negative: *Dilatris ixioides, Lacnanthes tinctoria, Shiekia orinocensis, Wachendorfia paniculata, Xiphidium floribundum.*

HALORAGIDACEAE
9 genera; 120 species

The family is partly aquatic and, although cosmopolitan, is concentrated in the southern hemisphere. Some are cultivated as aquarium plants.

The chemistry of the family is little known; there is one report of the presence of alkaloids in *Haloragis.* Of 15 species tested here one, *Myriophyllum propiniquum*, was positive.

Species negative for alkaloids included the following: *Gunnera macrophylla, G. perpensa, Gunnera sp., Haloragis acanthacarpa, H. chinensis, H. halconensis, H. micrantha, Laurembergia heterophylla, L. repens, L. tetandra, Myriophyllum humile, M. exalbescens, Prosperpinaca palustris. Gunnera* is now assigned a family of its own, Gunneraceae.

HAMAMELIDACEAE
28 genera; 90 species

This family is widespread but is found chiefly in the subtropics of southeast Asia. A few are North American, including the familiar witch hazel (*Hamamelis*).

Positive alkaloid tests have been recorded for a couple of genera in the family, but in the study reported here, 33 samples in 15 species gave only one positive result with a species already known to be alkaloidal: *Eustigma oblongifolium.*

Negative results were obtained with the following: *Corylopsis pauciflora, C. sinensis, C. spicata, C. veitchiana, Fothergilla gard-neri, Hamamelis japonica, H. virginiana, Liquidambar formosana, L. styracifolia, Ostrearia australiana, Parrotia persica, Trichocla-dus crinitus, T. grandiflorus, Trichocladus sp.*

HERNANDIACEAE
4 genera; 68 species

The Hernandiaceae are almost pantropical. Some are ornamental but otherwise economically unimportant.

Chemical investigation appears to have been confined to two of the genera, all samples of which have so far been found to contain alkaloids of the benzylisoquinoline type. As expected, the following were found positive in this study: *Gyrocarpus americanus* (formerly as *Gyrocarpaceae*), *Hernandia catalpifolia, H. cordigera, H. guianensis* (3/3), *H. jamaicensis, H. ovigera* (15/15), *H. papuana, H. peltata*, and one unidentified *Hernandia sp.*

Four species of *Sparattanthelium* were negative although aporphine alkaloids had been found in *S. uncigerum*; these negatives were *S. tupiniquinorum* and three unidentified species. The chemistry of the alkaloids has been reviewed.

HIMANTANDRACEAE
1 genus; 2 species

This family is found from eastern Malaysia to northern Australia. It is alkaloidal, nine out of nine samples of *Galbulimima balgraveana* were found to be positive.

HIPPOCASTANACEAE
2 genera; 15 species

This is a north temperate family extending to northern South America and with a few members in southeast Asia. We recognize the horse chestnut (*Aesculus hippocastanum*) and other *Aesculus* species as ornamentals.

Nitrogenous substances, but not alkaloids in the strict sense, are found in the family. Eight species (nine samples) were tested with one positive result: *Aesculus californica* (1/2).

Five other species of *Aesculus* were negative: *Aesculus assamia, A. hippocastaneum, A. parviflora, A. pavia, Aesculus sp.*

HUMIRIACEAE
8 genera; 50 species

The family name is sometimes spelled Houmiriaceae; it is tropical South American, occurring as far north as Costa Rica and with some representation in western Africa.

The family is not known for alkaloids; only one sample in this survey was positive: *Vantanea guianensis.* The remainder were negative: *Humiria balsamifera, H. floribunda, Humiria sp., Humiriastrum piraparanense, Saccoglottis guianensis, S. uchi, Saccoglottis sp., Vantanea contracta, V. cupularis, Vantanea spp.* (2).

HYDROCHARITACEAE
16 genera; 90 species

The family is cosmopolitan, found in warm fresh and salt waters of the world. Some are used as aquarium plants.

Alkaloids are not known. Tests of ten samples representing seven species yielded no positive result. The following were negative: *Elodea nuttallii, Hydrilla verticillata, Lagarosiphon major, L. muscoides, Limnobium spongia, Ottelia alismoides, Vallesneria americana.*

HYDROPHYLLACEAE
22 genera; 275 species

This family is almost cosmopolitan, found chiefly in the dry regions of western North America south to the Straits of Magellan with some representation elsewhere, except Australia.

There is a single report of the presence of alkaloids in *Phacelia* but tests on three species of the genus have been reported as alkaloid free. In the present study, no positive tests were noted. The following were negative: *Codon royenii, C. schenkii, Eriodictyon trichocalyx, E. crassifolium, Hydrolea spinosa, Nama biflora, N. densum, N. dichotomum, N. origanifolium, N. palmeri, N. parviflora, N. rupicolum, N. sericeum, N. subpetiolare, N. undulatum, N. dicholonum chasmogamum, Nemophila menziesii, N. fremontii, Phacelia heterophylla, P. humilis, P. imbricata, P. longipes, P. ramosissima, Phacelia sp., P. viscida, P. platycarpa, Pholistoma auritum, Wigandia kunthii, W. scorpioides.*

HYPOXIDACEAE
7 genera; 120 species

Lawrence has the members of this "family" in Amaryllidaceae, Willis in Hypoxidaceae, Cronquist (Mabberley) in Liliaceae. Two samples, *Curcilago orchioides* and *Hypoxis obtusa*, were negative.

I

ICACINACEAE
60 genera; 320 species

The Iacacinaceae constitute a tropical family with a few temperate representatives. Some are used for timber and others for local food and medicine.

The alkaloids of the family are of the emetine type (Wiegrebe, Kramer, and Shamma, 1984). Thirty-five samples constituting 27 species were tested with the following positive results: *Cassinopsis ilicifolia* (2/2, previously known), *C. tinifolia, Gastrolepis sp., Humiranthera rupestris* (1/3), *Medusanthera papuana* (now assigned to its own family, Medusandraceae), *Stemonurus papuanus, Urandra umbellata.*

Negative tests were given by *Apodytes dimidata, Citronella brassii, C. congonha, C. paniculata, Discophora guianensis, Emmotium fagifolium, E. nitens, E. nitidum, Emmotium sp., Humiranthera spinosa, H. duckei, Lophopyxis maingayi, Lophopyxis sp., Peripterigyium moluccanum, Polyporandra scandens, Poraqueiba guianensis, P.. paraensis, P. sericea, Pyrenacantha grandiflora, P. kamassana, Stemonurus sp.*

The position of *Peripterygium* is not clear; according to Mabberley, the genus is *Cardiopteris* in the family Cardiopteridaceae while Willis has it in Peripterygiaceae = Cardiopterygiaceae. A test of *P. moluccanum*, as indicated above, was negative.

REFERENCE

Wiegrebe, W., W. J. Kramer, and M. Shamma, *Journal of Natural Products 47* (1984) p. 397.

IRIDACEAE
92 genera; 1,800 species

This cosmopolitan family, represented especially well in South Africa, the eastern Mediterranean, and Central and South America, is known for its garden flowers, the spice saffron, and the genus *Iris* for its perfume oils.

Alkaloids in the usual sense are not found, but several genera yield tyramine derivatives and amino acids, some of which may be responsible for reports of alkaloids. Of 72 species, the following were positive in this study: *Aristea ecklonii, Ferraria antherosa, F. refracta, Gladiolus expersus, Homeria breyniana, Lapeirousia grandiflora, Nivenia stokoei, Sisyrinchium angustifolium* (1/2), *S. macrocephalum.*

Negative were *Anapalina nervosa, Aristea africana, A. bakeri, A. schizolaena, A. spiralis, A. thyrsiflora, Babiana falcata, B. patula, Bobartia indica, B. macrospatha, B. robusta, Crocosmia aurea, C. crocosmaflora, Cypella plumbea, Dietes sp., D. vegeta, Ferraria bechuanica, Gladiolus atropurpureus, G. melleri, G. multiflorus, G. paluster, Gladiolus sp., G. symmetranthus, G. villosus, Homeria pallida, Iris fosteriana, I. hertwegii, I. kaempferi, I. pumila, I. san-*

guinea, I. virginica, Ixia polystachya, Lapeirousia fissifolia, Lemmonia californica, Libertia pulchella, Micranthus tubulosus, Moraea moggii, M. polystachya, Moraea sp., M. spathulata, Neomarica coerulea, Orthosanthus sp., Patersonia fragilis, Romulea rosea, Sisryinchium bellum, S. exaltatum, S. nidulare, Sisyrinchium spp. (3), *S. striatum, Sparaxis grandiflora, Sphenostigma sellowiana, Tigrida sp., Tritonia crocata, T. crocosmiflora, Watsonia angusta, W. meriana, W. pillansii, W. pyramidata, W. spectabilis, W. transvaalensis.*

J

JUGLANDACEAE
7 genera; 59 species

The family has two areas of distribution: north temperate New World with extension through Central America and western South America to Argentina, and temperate Asia to Java and New Guinea. The trees are valued as timber and cabinet woods and, of course, for walnuts, hickory nuts, and pecans.

5-Hydroxytryptamine has been found in walnuts (*Juglans regia*); otherwise alkaloids are not known in the family. No positive tests were obtained for the following: *Carya glabra, C. ovata, C. pecan, C. tomentosa, Engelhardia acerifolia, E. colebrookiana, E. chrysolepis, E. rigida, E. formosana, Juglans allanthifolia, J. californica, J. cineria, J. mollis, J. nigra, J. regia, Pterocarya rhoifolia.*

JULIANACEAE
2 genera; 5 species

This family is found from Peru north to Central America. It has yielded some dyestuffs. Alkaloids are not known. A sample of *Amphipterygium adstringens*, the source of a red dye, was negative.

JUNCACEAE
10 genera; 325 species

A family of temperate and cold regions including mountainous areas of South America, the Juncaceae have their greatest diversity in the southern hemisphere. Some species are used for the weaving of mats and caning of chairs while others are ornamental.

Cyanogenesis has been noted in the family and there have been reports of "unnamed alkaloids," but none of the 23 samples representing 18 species gave a positive test for alkaloids: *Juncus acutus, J. brevicaudatus, J. bufonis, J. decipiens, J. exertus, J. leersii, J. lomatophyllus, J. maritimus, J. orthophyllus, J. papillosus, J. pallidus, J. pelocarpus, J. setchuensis, J. setchuensis* var. *effusoides, Juncus sp., Luzula effusa, Prionium serratum, Xerotes sp.* (= *Lomandra,* now in Xanthorrhoeaceae).

JUNCAGINACEAE
4 genera; 18 species

This is a small family of herbs found in the wet areas of temperate and cold regions. Some species of *Triglochin* are edible; others are toxic due to cyanogenesis, common in the family. *Triglochin maritima* and *T. striatum* were alkaloid negative.

L

LABIATAE
221 genera; 5,600 species

The mint family is sometimes called the Lamiaceae. It consists chiefly of cosmopolitan herbs prominent in the area from the Mediterranean to central Asia. The mints have long been used in medicine, as spices, and in some perfumes. They are familiar kitchen herbs (mint, basil, oregano, etc.). Many are cultivated as ornamentals.

Sesquiterpenoid alkaloids are found in some members of the family, stachydrine and many pseudo- or protoalkaloids in others.

Six hundred and twenty-two samples representing 405 species

were tested; three of those were known to be alkaloidal from earlier reports: *Marrubium vulgare* (1/5), *Nepeta cataria*, *Rosmarinus officinalis* (1/3).

Other positives included *Acrocephalus sp.* (1/4), *Aeolanthus parvifolius*, *Anisomeles indica* (2/3), *Ballota africana* (2/2), *Cedronella sp.*, *Conradina canescens*, *Dysophylla* (= *Pogostemon*) *cruciata*, *Elsholtzia stauntonia*, *Endostemon tereticaulis* (1/2), *Hedeoma pulegeoides*, *Hoslundia opposita* (3/4), *Hyptis subtilis*, *H. uliginosa*, *H. verticillata*, *Iboza* (= *Tetradenia*) *brevispicata* (1/2), *I. riparia* (1/3), *Lasiocorys* (= *Leucas*) *capensis* (1/2), *Leonotis leonurus* (2/2), *L. leonitis*, *L. melleri*, *L. nepetaefolia* (3/8), *L. sibiricus* (1/7), *Leucas sp.*, *Melissa parviflora* (1/4), *Mentha alopecuroides*, *Monarda punctata* (1/2), *Orthosiphon rubicundus*, *Perilla frutescens*, *Plectranthus calycinus*, *Pogostemon parviflorus*, *Pycnanthemum virginianum* (1/2), *P. incanum*, *Salazaria mexicana*, *Salvia apiana* (1/2), *S. azurea*, *S. ballotaeflora* (1/4), *S. breviflora* (1/2), *S. chameloegana* (1/2), *S. clandestina*, *S. connivens*, *S. kerlii*, *S. leucophylla*, *S. namaensis*, *S. polystachya* (1/7), *S. runcinata*, *S. sessei* (1/3), *Scutellaria incana* (1/2), *S. pauciflora*, *Stachys sp.* aff. *bigelovii*, *S. albens*, *S. burchelli*, *S. nigricans*, *S. rigida* (1/2), *S. thunbergii*, *Tetraclea coulteri*, *Teucrium canadense* (2/3), *T. cubense* (2/2), *Tinnea sp.*, *Trichostema parishii*.

The following were negative: *Acrocephalus indicus*, *Aeolanthus katangensis*, *A. rehmannii*, *Agastache barbeii*, *A. mexicana*, *A. rugosa*, *Ajuga macrosperma*, *A. dichtyocarpa*, *Alvesia rosmarinifolia*, *Becium angustifolium*, *B. burchellianum*, *B. homblei*, *B. kynanum*, *B. obovatum*, *Brunella vulgaris*, *Cedronella auriantiaca*, *C. mexicana*, *C. contracta*, *Clinopodium coccineum*, *C. colbrookea*, *C. confinis*, *C. dantatum*, *C. oppositifolium*, *Coleus aromaticus*, *C. amboinicus*, *C. barbatus*, *C. latifolius*, *C. neochilus*, *C. rehmannii*, *Coleus sp.*, *Collinsonia anisata*, *C. canadensis*, *Craniotome versicolor*, *Cunila galioides*, *C. godioides*, *C. lythrifolia*, *C. origanoides*, *C. pyncantha*, *Elsholtzia patrini*, *Endostemon obtusifolius*, *E. tenuiflorus*, *Engelarastrum djalonense*, *Gardongia mexicana*, *Geniosporum angolense*, *G. paludosum*, *Glechoma hederacea*, *Hedeoma drummondii*, *H. palmeri*, *H. patens*, *Hemizygia bracteosa*, *H. elliotii*, *H. latidens*, *H obermeyerae*, *H. petiolata*, *H. rehmannii*, *Hemizygia sp.*, *H. thorncroftii*, *Hesperozygis myrtoides*, *Hoslundia verti-*

cillata, Hoxis capitata, Hyptis albida, H. asperrima, H. capitata, H. complicata, H. crinita, H. ditassoides, H. emotyi, H. gaudichaudii, H. interrupta, H. lappacea, H. laxiflora, H. lutescens, H. marifolia, H. mutabilis, H. paraensis, H. nervosa, H. nudicaulis, H. pectinata, H. plectranthoides, H. radiata, H. recurvata, H. rhomboidea, H. rhytidea, Hyptis spp. (15), *H. spicata, H. spicigera, H. suaveolens, H. umbrosa, H. urticoides, H. vestita, H. villosa, Iboza galpinii, Lamium amplexicaule, Leonotis dysophylla, L. microphylla, L. mullissima, Leonotis sp., Leonurus sikinicus, Lepechinia nelsonii, Lepechinia speciosa, L. spicata, Leucas aspera, L. decemdentata, L. glabrata, L. lanata* var. *candida, L. martinicensis, L. milanjiana, L. neufliseana, Lycopus spp.* (2), *Marsypianthes chamaedrys, Mentha aquatica, M. arvensis, M. longifolia, M. memorosum, M. piperita, M. pulegium, M. rotundifolium, Mesona procumbens, Micromeria biflora, Monarda austromontana, M. didyma, M. fistulosa, Monardella villosa, Nauthochilus labiatus, Nepetoides sp., Ocimum americanum, O. basilicum, O. fruticolosum, O. gratissimum, O. micranthum, O. sanctum, O. sellowii, Ocimum spp.* (2), *O. suave, O. urticifolium, Orthodon formosanum, O. lanceolatum, Orthosiphon aristatus, O. tubiformis, Peltodon longipes, P. radicans, P. rugosus, Perilla sp., Phyllestegia grandiflora, Plectranthus calycinus, P. coetsa, P. grandidentatus, P. cylindraceus, P. dolichopodus, P. eckloni, P. fruticosus, P. hirtus, P. hoslundioides, P. laxiflorus, P. myrianthus, P. nummularis, P. oncanus, P. saccatus, P. sanguineus, P. swynnertonii, P. tomentosa, Plectranthus sp., Pogostemon cablin, Poliomintha glabrescens, P. longiflora, Poliomintha sp., Prostanthera melissifolia, P. striatiflora, P. euphrasioides, Prunella vulgaris, Pseudocunila sp., P. montana, Pycnanthemum incanum, Pycnostachys dewildmaniana, P. kassneri, P. reticulata, P. urticifolia, Raphiodon echinus, Rhabdocaulon gracilis, R. lavenduloides, Rhabdocaulon spp.* (2), *R. villosa, Salvia africana coerulea, S. africana lutea, S. amarissima, S. apiana, S. azurea, S. cardinalis, S. chamadedryoides, S. cimabarina, S. clevelandii, S. coccinea, S. columbariae, S. concolor, S. cuneifolia, S. curviflora, S. dolomitica, S. dryophila, S. elegans, S. emasiata, S. excelsa, S. filipes, S. gesneraeflora, S. glechomaefolia, S. gracilis, S. greggii, S. guaranitica, S. hauinulus, S. hayatana, S. helianthemifolia, S. hirsuta, S. hispanica,*

*S. hyptoides, S. iodantha, S. isochroma, S. karwinskii, S. lachnosta-
chys, S. laevis, S. langlassei, S. lasiantha, S. lasiocephala, S. lavan-
duloides, S. leptophylla, S. leptostachys, S. leucantha, S. longistyla,
S. lyrata, S. mandrensis, S. melissodora, S. mellifera, S. merlii,
S. mexicana, S. mexicana minor, S. mexicana neurepia, S. microphylla, S. microphylla neurepia, S. mocinoi, S. nana, S. nipponica
formosana, S. pachyphylla, S. paranensis, S. patens, S. perblanda,
Salvia ap. aff. polystachya, S. pruinosa, S. prunelloides, S. puberula, S. pulchella, S. purpurea, S. reflexa, S. regla, S. riparia,
S. rubiginosa, S. sanctaeluciae, S. scapiformis, S. setulosa, S. sisym-
brifolia, S. spathacea, Salvia spp. (18), S. stricta, S. thyrsiflora,
S. tiliaefolia, S. triangularis, S. unicostata, S. uruapana, S. veronicaefolia, Satureja chinensis parviflora, S. macrostemma, S. rigida,
Scutellaria austinae, S. coerulea, S. incana, S. laterifolia, S. potosi-
na, S. racemosa, S. repens, Soldia regia, Stachys aethiopica,
S. agraria, S. ajugoides, S. bigelovii, S. bullata, S. coccinea,
S. drummondii, S. grandifolia, S. integrifolia, S. multiflora, S. nepe-
taefolia, S. pringlei, S. repens, S. rugosa, Stachys sp., Suzukia shiki-
kunensis, Synclostemon densiflorus, S. argenteus, S. macrophyllus,
Teucrium racemosum, T. africanum, T. integrifolium, T. visdidum,
Tinnea juttea, T. vestita, T. zambesiaca, Trichostema dichotomum,
T. lanatum, T. lanceolatum, T. setaceum.*

LACISTEMATACEAE
1 genus; 14 species

This is a small group of tropical American trees and shrubs. Little
is known of the chemistry of the family; alkaloids have not been
reported.

No positive tests were obtained with *Lacistema aggregatum,
L. pubescens*, two unidentified *Lacistema spp.*, and one of *Lozania*.

LARDIZABALACEAE
8 genera; 21 species

The family extends from Japan and China to the Himalayas with
two genera in Chile. Some fruits are edible but have achieved no
economic importance.

Alkaloids are not known. Tests conducted in this study on *Akebia longeracemosa*, *A. quinata,* and *Stauntonia hexaphylla* were negative.

LAURACEAE
45 genera; 2,200 species

Most representatives of the family are found in tropical southeast Asia with a few in Africa and the Mediterranean area. The family enjoys economic importance for its aromatic oils used as spices, medicinals, flavors, and odors (cinnamon, camphor, sassafras, etc.) and for the familiar fruit, avocado.

Alkaloids are found throughout the family and reviews of their occurrence and structures are available (Guinandeau, LeBoeuf, and Cave, 1975; Kametani, Ihara, and Honda, 1976). In this study, 371 samples representing 177 species were tested. Several of them had been reported as alkaloidal earlier: *Alseodaphne archboldiana*, *Cassytha filiformis* (9/14), *C. laubatii* (3/4), *Cryptocarya cinnamomifolia*, *C. chinensis* (3/3), *C. erythroxylon, Laurus nobilis, Lindera membranacea* (2/2), *L. umbellata, Litsea cubeba* (2/3), *L. glutinosa* (2/3), *L. reticulata, Neolitsea acuminatissima* (3/5), *Ocotea* (= *Nectandra*) *rodioei, Phoebe clemensii, Sassafras albidum.*

Other alkaloid-positive samples included: *Actinodaphne natoensis* (1/3), *A. tomentosa, Aiouea schomburgkii* (2/2), *Aniba cylindrifolia, A. fragrans, A. parviflora, A. roseodora, Beilschmiedia spp.* (4/5), *B. grammiena* (bark), *Cassytha americana, C. ciliolata, C. glabella, C. pubescens, C. racemosa, Cinnamomum camphora* (1/5), *C. loureirei, Cinnamomum spp.* (2/2), *Cryptocarya foveolata, C. latifolia, C. pauciflora, Cryptocarya spp.* (5/6), *C. saligna* (1/2), *Dehaasia novoguineensis, Embuia sp., Endlichera tschudiana, Licaria americana, L. camera, Licaria sp.* (1/2), *Lindera benzoin* (3/3), *Litsea akonensis, L. domarensis, L. glaucescens* (1/5), *Litsea spp.* (8/11), *Nectandra globosa, N. sinuata, Nectandra spp.* (8/14), *N. superba, N. surinamensis, Neolitsea acutotrinervia* (1/2), *N. cassia, N. dealbata, N. kotoensis, Neolitsea sp., Ocotea barcellensis, O. glandulosa, O. hypoglauca, O. macropoda* (2/2), *O. rubra, O. simulans, O. schomburgkiana, O. spixana, Ocotea spp.* (6/33), *Persea ramissonis, P. rufotomentosa* (1/2), *P. fuliginosa, Persea*

spp. (2/5), *Phoebe attenuata, P. trianae, P. tampicensis, Umbellula-ria californica* (1/2).
The following were negative: *Acrodiclidium aureum, A. puchury, Actinodaphne mushaensis, A. morrisonense, A. pedicillata, Aiouea densiflora, Aniba burchelli, A. firmula, A. guianensis, A. hostman-niana, A. ovalifolia, A. panurensis, Aniba sp., Beilschmiedia mexi-cana, B. rohliana, Beilschmiedia sp., Benzoin glaucum, Borbonia* (= *Persea*) *cordata, B. lanceolata, Cinnamomum baileyanum, C. englerianum, Cinnamomum sp.* aff. *gilgianum, C. insulari-mon-tanum, C. japonicum, C. micranthum, C. obtusifolium, C. quadran-gulum, C. sellowianum, C. virens, C. zeylanicum, Cryptocarya lieb-ertiana, C. mackinnoniana, C. moschata, C. oblata, C. wyliei, Endriandra glauca, E. grandiflora, E. sieberi, Endlicheria arunci-flora, Endlicheria sp.* aff. *arunciflora, E. longicaudata, E. panicu-lata, Laurus esoinosa, L. capitata, L. communis, Licaria coriacea, L. mahuba, Licaria sp., Machilus arisaensis, M. thunbergii, Nec-tandra coriacea, N. dioica, N. glabrescens, N. lanceolata, N. lucida, N. rectinervia, N. reticulata, N. rigida, N. sanguinea, Neocinnamo-mum delavayi, Neolitsea konishii, N. parvigemma, N. variabillima, Ocotea abbreviata, O. aciphylla, O. acutifolia, O. bullata, O. canal-iculata, O. carabobensis, O. corymbosa, O. ensifolia, O. glaziovii, O. glomerata, O. guianensis, O. notata, O. opaca, O. opifera, O. porosa, Ocotea sp.* aff. *preciosa, O. rubiginosa, O. tristis, O. viridis, Persea americana, P. cordata, P. gratissima, P. palustris, P. venosa, Phoebe ehrenbergii, P. formosana, Phoebe sp., Sassafras sp., Sylvia itatuba, Systemonodaphne sp., Urbanodendron verrucosum.*

REFERENCES

Guinandeau, H., M. LeBoeuf, and A. Cave, *Lloydia 38* (1975) p. 275.
Kametani. T., M. Ihara, and T. Honda, *Heterocycles 4* (1976) p. 483.

LECYTHIDACEAE
20 genera; 280 species

The trees of this family, found especially in the rain forests of South America, yield timber and the familiar Brazil nuts.

Alkaloids from the genus *Couroupita* have been described. Three other genera gave positive tests in the survey reported here: *Eschweilera jarana, E. ucayaliensis, Holopyxidum* (= *Lecythis*) *jarana, Lecythis paraensis* (2/2).

Negative results were obtained with the following: *Allantoma lineata, Barringtonia acutangula, B. asiatica, B. edulis, B. gracilis, B. racemosa, Barringtonia spp.* (3), *Cariniana estrellensis, Couratari pulchra, Couratari sp., Courouptia guianensis, Eschweilera amazonica, Eschweilera sp.* aff. *odora, Eschweilera spp.* (2), *Gustavia hexapetala, Gustavia spp.* (2), *Lecythis zabucajo, Napoleona imperialis, Planchonia papuana.*

LEGUMINOSAE
657 genera; 16,400 species

This is one of the largest families of flowering plants, cosmopolitan in distribution, and with many species introduced in several parts of the world. Three subfamilies are considered as separate families by some taxonomists.

Economically the family is important as a major source of food for the world: peas, beans, forage, as well as gums, resins, and oils, but several species are toxic to grazing animals. The literature of the family is vast.

Reviews of the chemistry of the family have been published (Arora, 1983; Salatino and Gottlieb, 1981). Alkaloids are common throughout. In the study reported here some 2,600 samples that included 1,636 species were tested to give a total of 353 positives. Many of these were previously known: *Abrus praecatorius, Acacia complanata, A. concinna, A. harpophylla, A. kettlewelliae, A. longifolia, A. nerifolia, A. polystachya, A. senegal, A. verticillata* (2/2), *Aeschynomene indica, A. julibrissin* (1/3), *Argyrolobium megarhizum, Baptisia alba, B. lanceolata, B. psammnophila, B. simplicifolia, B. tinctoria* (3/3), *Calliandra portoricensis* (2/3), *Calpurnia rosea* (1/5), *Cassia absus, C. carnaval, C. excelsa, C. laevigata* (2/6), *C. leptocarpa* (1/2), *C. siamea, Crotalaria agatiflora, C. alata, C. anagyroides* (2/2), *C. incana, C. verrucosa* (2/4), *Cytisus scoparia* (3/3), *Dalea tuberculata* (2/6), *Dolichos lablab, Entada phaseoloides, Erythrina abyssinica* (2/2), *E. americana* (3/4),

E. berteroana, E. breviflora, E. chiapasensis, E. cristagalli (1/2),
E. flabelliformis (2/2), *E. fusca, E. herbacea* (2/2), *E. occidentalis,*
E. poeppigiana, Erythrophloem chlorostachya (2/6), *E. guineense,*
Gleditsia triancanthos (2/6), *Hovea heterophylla, H. longifolia*
(4/4), *H. longipes* (15/15), *H. trisperma, Indigofera endecaphylla*
(1/3), *I. suffruticosa* (1/8), *Lupinus andersonii* (2/2), *L. arboreus*
(3/3), *L. elegans, L. excubitus, L. nanus, L. perennis, L. succulentus,*
L. villosus, L. westianus, Melilotus alba (1/2), *M.* indica (1/6),
Ormosia emarginata, Parkinsonia aculeata (1/4), *Petalostylis labi-*
cheoides, Pithecellobium flexicaule (2/2), *P. saman* (3/3), *Podalyria*
calyptrata, Prosopis juliflora (9/10), *Samanea saman, Sophora to-*
mentosa (4/4), *S. angustifolia, S. flavescens* (2/2), *S. secundiflora*
(3/3), *S. tetraptera* (3/3), *Spartium junceum, Templetonia retusa,*
Tephrosia candida (1/2).

Many other samples were likewise positive: *Abrus cantonensis,*
A. fruticosus, Acacia adunca, A. aneura, A. anthelmintica, A. ar-
gentea, A. armata, A. chaconesis, A. cowliana, A. erubescens (1/2),
A. ferruginea, A. giraffae, A. hebeclada (1/2), *A. holscerica,*
A. intsia, A. leptocarpa, A. lutea (1/2), *A. mackothyrsa* (1/2),
A. mellifera (3/3), *A. nigrescens, A. nitidifolia, A. polycantha* (2/2),
A. schweinfurthii, A. sophorae, Acrocarpus fraxinifolius, Aeschyno-
mene bracteosa, Albizzia brevifolia, A. zygia, A. antunesiana, Am-
phitalea intermedia, Aotus subglauca, Argyrolobium transvaalense,
A. shirense, Aspalathus lactea, A. quinquefolia, A. spinosa (3/3),
Aspalathus spp. (4/9), *Astragalus rosmarinaefolius, Ateleia ptero-*
carpa, Bauhinia garipepensis (2/2), *B. thonningii, Bolusanthus spe-*
ciosus (2/2), *Brongniartia mollis, B. glabrata* (2/2), *B. intermedia*
(3/3), *B. parryi, Butea monosperma, Caesalpinia crista, C. ferrea,*
Calpurnia floribunda, C. intrusa, C. subdecandra, Campisandra
laurifolia (1/2), *Canavalia ensiformis* (1/4), *C. maritima* (2/4),
C. lineata, Cassia adiantifolia (2/4), *C. alata* (1/7), *C. floribunda,*
C. goldmannii (1/2), *C. javanica, C. lucens, C. mimosoides* (1/6),
C. petersiana (1/2), *C. polyantha, C. shinneri, Cassia spp.* (6/53),
C. spectabilis (3/3), *Chamaecrista* (= *Cassia) fasciculata, Clitoria*
sp. aff. *arborea, C. rubiginosa* (1/2), *C. terneata* (2/6), *Coelidium*
fourcadei, C. spinosum, Colvillea racemosa (1/2), *Copaifera bau-*
miana, Crotalaria capensis (1/2), *C. comosa, C. doidgeae, C. lan-*
ceolata (2/4), *C. natalensis, C. natalitia* (1/2), *C. nubica, C. ptero-*

caule, C. saltiana (2/4), *Crotalaria spp.* (6/20), *C. striata, C. teretifolia, C. vitellina, Crudia oblonga, Cymbosema roseus, Cytisus sp., Dalbergia arbutiflora, D. ecastophyllum* (1/4), *Dalbergia sp.* (1/6), *D. violacea* (fruit), *Dalea californica* (1/2), *Daviesia mimosoides, D. ulicina, Derris spp.* indet. (4/8), *D. urucu, Desmodium spp.* (3/13), *D. uncinatum* (1/2), *Dichrostachys glomerata* (1/3), *D. nyasana* (1/3), *Dimorphandra mollis, Dolichos sp., Dussia sp., Ebenopsis* (= *Pithecellobium*) *flexicaulis* (2/2), *Elephantorrhiza goetzii* (1/2), *Entada sp., Eriosema ellipticum, E. glomeratum* (2/2), *E. psoraloides, Erythrina indica, E. humeana, E. lysistemon* (2/2), *E. stricta, Erythrina spp.* (5/6), *Gompholobium virgatum, Guibourtia coelospermum* (2/3), *Guilandina* (= *Caesalpinia*) *crista, Hovea acutifolia, H. chorizemifolia, Hymenolobium sp., Indigofera erecta* (1/3), *I. australis, I. heterotricha* (1/3), *I. sanguinea, Indigofera sp.* (1/6), *I. sumatrana* (1/2), *Lathyrus latifolius* (1/4), *Lebeckia cystoides* (1/2), *L. macrantha, L. plukenetiana* (3/3), *L. pungens, L. spinescens, Lessertia macrostachya, Leucaena macrocarpa, Lonchocarpus constrictus, Lonchocarpus sp., Lotononis umbellata, Lotononis sp., Lotus chihuahuana, L. crassifolius, Lupinus alpestris, L. chamissonis, L. confertus, L. cumulicola, L. densiflorus, L. ehrenbergeri, L. giganteus, L. grayi, L. hirsutissimus, L. leucophyllus, L. longifolius, L. luteus* (2/2), *L. sellulus* (2/2), *Lupinus spp.* (16/18), *L. sparsiflorus, L. squamaecaulis, L. stiversii, L. submontanus, L. uncinatus* (2/2), *Machaerium macrophyllum, M. multifoliatum, Machaerium sp.* (1/6), *Macrolobium acaciaefolium, M. arenaria, Molololobium adenodes* (1/2), *M. calycinum, M. exudans, M. macrocalyx, Medicago hispida* (1/2), *M. polymorpha* (1/2), *Milletia sutherlandii, M. stuhlmanii, Mimosa pudica* (2/6), *Mirbelia dilatata, Moldenhowera blanchettii* (1/4), *Mora paraensis, Mundulea sericea* (3/4), *Neptunia oleracea, Ormosia amazonica* var. *venenifera* (seed), *O. coutinhoi* (2/2), *O. microcalyx, Ormosia sp.* (1/4), *O. toledoana, Oxylobium ilicifolium, Parkia sp., Parochetus communis, Pearsonia aristata, P. filifolia* (1/2), *Pearsonia sp.* (1/2), *Pentaclethra macrophylla, Phaenohoffmania* (= *Pearsonia*) *cajanifolia, Phaseolus lathyroides* (1/2), *Phaseolus sp.* (1/7), *P. vulgaris, Phyllodium pulchellum, Pithecellobium* aff. *dumosum, P. arboreum* (1/2), *P. berterianum, P. ciricinnale, P. cochleanum* (1/2), *P. jupunba, P. lanceolatum* (1/2), *P. langsdorfii* (2/2), *P. leptophyllum* (1/2),

P. racemosum, Platymiscium polystachum, Pleiospora (= *Phaeno-hoffmania*) *cajanifolia* (2/2), *P. lateobracteoplata, Podalyria cunei-folia, P. glauca, P. myrtillifolia, Podalyria sp.*, *Podopetalum* (= *Ormosia*) *ormondii* (1/3), *Poecilanthe effusa, Priestleya hirsuta, Prosopis insularum, P. palmeri, Psoralea aculeata, P. oligophylla, P. polysticta, Pterocarpus rohrii, Pterogyne nitens, Pultenea hart-manii, P. subternata, P. villosa, Pycnospora luteola, Rafnia opposi-ta, R. ovata, Rhynchosia albissima* (1/2), *R. monophylla, R. nitens, R. pauciflora, Sophora vicifolia, Styphnodendron coriaceum, Su-therlandia frutescens, Swainsonia burkittii, Swartzia simplex* var. *grandiflora, Swartzia sp.* (1/3), *Sweetia elegans, S. dasycarpa, Sweetia sp., Tachigalia paniculata* (1/3), *Tachigalia sp., Templeto-nia egena, Tephrosia adunca* (1/3), *T. astrogaloides, T. forbesii, T. paniculata* (2/2), *T. polystachyoides, Tephrosia sp., T. sinapou, T. toxicaria, T. tzaneensis, T. vogelii, Trifolium burchellianum, T. carolinianum, T. hybridum* (2/3), *T. pratense* (1/4), *Vicia villosa* (1/3), *Vigna nuda, Vigna sp., Virgilia divaricata, V. oroboides, Wil-borgia obcordata, Wisteria floribunda, Zornia diphylla* (1/4).

Negative tests were also common: *Acacia actalensis, A. aculae-tissima, A. angustissima, A. anisophylla, A. ataxacantha, A. berlan-deria, A. bilimekii, A. borleae, A. brandegiana, A. caffra, A. confu-sa, A. constricta, A. cornigera, A. coulteri, A. crassifolia, A. cymbispina, A. davyi, A. decurrens, A. dolichastodya, A. farne-siana, A. filicicoides, A. gerrardii, A. gillettiae, A. glomerata, A. greggii, A. haematoxylon, A. hilliana, A. jacquemontia, A. junci-folia, A. karroo, A. kempeana, A. koa, A. leichhardtii, A. macrantha, A. maidenii, A. melanoxylon, A. mitchellii, A. nilotica, A. panicula-ta, A. parviflora, A. patens, A. pennatula, A. permixta, A. pineto-rum, A. polyphylla, A. rehmannii, A. retinoides, A. retivemia, A. robusta, A. roemeriana, A. sieberiana, A. simplicifolia, Acacia sp., A. spirorbis, A. sauveolens, A. subangulata, A. tenuispina, A. tetragonophylla, A. tortilis, A. tortuosa, A. unijuga, A. villosa, A. wrightii, A. xanthophloea, Adenanthera abrosperma, A. pavoni-na, Adesmia ciliata, A. paraensis, Adesmia spp.* (2), *Aeschynomene abyssinica, A. amorphoides, A. aspera, A. brasiliana, A. mimosifo-lia, A. nyassana, A. rehmannii, A. schimperi, A. sellowii, A. sensiti-va, A. trigonocarpa, Affonsea edwallii, Afzelia bijuga, A. cuan-zensis, Anicia zygomeris, Albizia adianthifolia, A. amara,*

A. anthelmintica, A. coriaria, A. distachya, A. falcataria, A. forbesii, A. fulva, A. harveyi, A. lebbeck, A. occidentalis, A. odoratissima, A. peretsiana, A. procera, Albizia spp. (2), *A. tanganyicensis, A. thozetiana, A. veriscolor, Aldina sp., Alexa bauhiniflora, Alisycarpus benguamensis, A. glumaceus, A. longifolius, A. nummularifolius, A. rugosus, A. vaginalis, A. zeyheri, Amherstia nobilis, Amorpha canescens, A. croceolanata, A. frutescens, Amphicarpa bracteata, Andira fraxinifolia, A. humilis, A. inermis, A. nudipes, A. retusa, Andira spp.* (3), *Apoplanesia paniculata, Apuleia molaris, Arachis hypogea, Archidendron lucyi, Argyrolobium collinum, A. tomentosum, A. wilmsii, Ascidia piscipula, Aspalathus ciliaris, A. contaminatus, A. ericifolia, A. suffruticosa, Astragalus crotalariae, A. curvicarpus, A. esperanzae, A. humboldti, A. leucopsis, A. leptocarpus, A. lentiginosus, A. mollismus, A. seatoni, A. sinicus, A. strigulosus, A. tolucanus, A. wootoni, A. scarabaeoides, Baphia bequaertii, B. obovata, Baptisia australis, Batesia floribunda, Bauhinia brachycalyx, B. championi, B. divaricata, B. excisa, B. forficata, B. galpinii, B. glabra, B. glauca, B. hookeri, B. kirkii, B. longifolia, B. macrostachya, B. macrantha, B. manca, B. mendoncae, B. mexicana, B. monandra, B. pes-captae, B. racemosa, B. ramosissima, B. rufa, Bauhinia spp.* (5), *B. spathacea, B. splendens, B. tancalosa, B. tomentosa, B. ungulata, B. vahlii, B. variegata, B. winitii, Bossiaea cinerea, B. microphylla, B. rhombifolia, B. scontechnii, Bowdichia brasiliensis, Bowdichia sp., B. virgiloides, Brachystegia boehmii, Brachystegia sp., Brownea grandiceps, B. ariza, B. latifolia, B. longipedicelata, B. macrophylla, Brownea spp.* (3), *Burkea africana, Butea superba, Caesalpinia jayabo, C. bonducella, C. coriaria, C. crista, C. cacalaco, C. eriostachys, C. ferrea, C. peltophoroides, C. platyloba, C. pulcherrima, C. pyramidialis, C. rubicunda, Caesalpinia spp.* (4), *C. vernalis, Cajanus cajan, Calliandra angustifolia, C. bracteosa, C. californica, C. cryophylla, C. decrescens, C. emarginata, C. eriophylla, C. grandiflora, C. houstoniana, C. inaequilatera, C. lagunae, C. parviflora, C. rotundifolia, Calliandra spp.* (7), *C. surinamensis, C. tenuiflora, C. tweedii, Calopogonium coeruleum, C. muconoides, C. sericeum, Campisandra laurifolia, Camptosema sp., Canavalia angustifolia, C. bonariensis, C. gladiata, C. lineata, C. microcarpa, C. obtusifolia, C. plagiosperma, C. villo-*

sa, *Cantharospermum scarabaeoides, Carmichaelia australis, Cassia abbreviata, C. apocouita, Cassia sp.* aff. *apocouita, C. angulata, C. armata, C. auriculata, C. australis, C. bauhinioides, C. bicapsularis, C. biflora, C. brachypoda, C. cana, C. cathartica, C. catingae, C. choriophylla, C. colutedides, C. decubens, C. densiflora, C. dentata, C. desolata, C. desvauxii, C. diphylla, C. fasciculata, C. ferruginea, C. fistula, C. flexuosa, C. glauca, C. glutinosa, C. gracilis, C. gracilior, C. grandis, C. grandistipula, C. greggii, C. hebecarpa, C. hirsuta, C. hispidula, C. hoffmannseggii, C. holwayana, C. hypoleuca, C. italica, C. multijuga, C. katangaensis, C. lanceolata, C. latistipulata, C. langsdorfii, C. leiophylla, C. leschenaultiana, C. macdougaliana, C. macrophylla, C. mucronata, C. nicitans, C. nigra, C. occidentalis, C. ochnacea, C. oligophylla, C. patellaria, C. potosina, C. potentilla, C. pringlei, C. pubescens, C. punctata, C. pudibunda, C. quarrei, C. reinformis, C. repens, C. reticulata, C. rotundata, C. roemeriana, C. rotundifolia, C. scleroxylon, C. sanguinea, C. sophora, C. striata, C. tomentosa, C. tora, C. undulata, C. uniflora, C. venticulata, C. vogeliana, Catalpa bignonioides, Catanus indicus, Centrolobium paraense, C. tomentosum, Centrosema bracteosum, C. plumieri, C. pubescens, Centrosema spp.* (2), *C. triquetrum, C. virginianum, C. pubescens, Ceratonia siliqua, Cercidium floridum, C. praecox, C. texanum, Cercis canadensis, C. occidentalis, C. siliquastrum, Chaetocalyx brasiliensis, Chamaecrista aspera, C. brachiata, C. multipinnata, Chytroma valida, Cicer arctinum, Cladrastis lutea, Clathrotropis brachypetala, Clitoria arborescens, C. arborea, C. javitensis, C. leptostachya, C. lineata, Clitoria sp., Collaea neesii, C. speciosa, Cologonia mortia, Colophospermum mopane, Colutia frutescens, Conzattia multiflora, Copaifera duckei, C. langsdorfii, C. martii, C. reticulata, C. trapezifolia, Coronilla varia, Coursetia arborea, Coumarouna odorata, Craibia brevicaudata, Cratylia floribunda, Crotalaria aculeata, C. annua, C. anthyllopsis, C. aridicola, C. brownei, C. burkeana, C. calcycina, C. cephalotes, C. cylindrostachys, C. ferruginea, C. filicaulis, C. florida, C. glauca, C. goreensis, C. hislopii, C. juncea, C. kapirensis, C. kirkii, C. kipandensis, C. lachnophora, C. linifolia, C. longirostrata, C. microcarpa, C. monteiroi, C. mollicula, C. mucronata, C. natalitia, C. nicholsonii, C. onodoides, C. platycephala, C. podocarpa,*

C. pohliana, C. prostrata, C. pumila, C. quinquifolia, C. recta, C. reptans, C. retusa, C. rogersii, C. saggittalis, C. schiediana, C. sericea, C. senegalensis, C. sericifolia, C. similis, C. spectabilis, C. stipularis, C. strehlowii, C. trifoliastrum, C. usaramensis, C. virgulata, Crudia amazonica, C. parivoa, C. tomentosa, Cryptosepalum pseudotaxus, Cyclopia genistoides, Cyclopia sp., C. subternata, Cymbosema roseum, Cynometra sp., Dahlstedtia pinnata, Dalbergia armata, D. brasiliensis, D. densa, D. glandulosa, D. hancei, D. lanceolata, D. millettii, D. monetaria, D. nitidula, D. obovata, D. sisso, D. sericea, D. variabilis, Dalbergiella nyassae, Dalea acutifolia, D. alopecuroides, D. arborescens, D. citriodora, D. crassifolia, D. eysenhardtioides, D. formosa, D. fremontii, D. frutescens, D. humilis, D. inconspicua, D. lasiostachya, D. leucostoma, D. lozanoi, D. mollissima, D. nutans, D. pogonothera, D. polyadenia, D. psoraleoides, D. schottii, D. seemanni, D. sericea, Dalea spp. (2), D. submontana, D. triphylla, D. uncifera, Daubentonia punicea, Daviesia brevifolia, D. corymbosa, D. latifolia, D. wyattiana, Delonix regia, Derris elegans, D. elliptica, D. floribunda, D. glabrescens, D. pterocarpa, D. rariflora, D. trifoliata, Desmanthus virgatus, Desmodium abscendens, D. amplifolium, D. asperum, D. axillare, D. barbatum, D. buergeri, D. caffrum, D. canadense, D. canum, D. ciliare, D. cinerium, D. discolor, D. dispersum, D. elegans, D. gyrans, D. helenae, D. heterocarpum, D. heterophyllum, D. hirtum, D. laxiflorum, D. laxus, D. microphyllum, D. orbiculare, D. ovalifolium, D. paniculatum, D. perplexum, D. plicatum, D. polycarpum, D. pringlei, D. purpureum, D. repandum, D. salicifolium, D. scorpiurus, D. sequax, D. sequax sinuatum, D. spirale, D. styracifolium, D. tanganyikense, D. triflorum, D. tortuosum, D. triquetrum, D. umbellatum, D. variegatum, D. venustrum, Detarium senegalense, Dialium guianense, D. schlechteri, Dichrostachys glomerata, D. cinerea, Dillwynia floribunda, D. glaberrima, D. retorta, D. sericea, Dimorphandra multiflora, D. parviflora, Dimorphandra sp., Dinizia excelsa, Dioclea excelsa, D. erecta, D. guianensis, D. macrocarpa, D. megacarpa, D. reflexa, Dioclea sp., D. violacea, D. virgata, Diphysa robinioides, Diplotropis martiusii, D. purpurea, Diptrix nudipes, Dolichos daltonii, D. eriocaulis, D. gibbosus, D. gululu, D. lignosus, D. malosanus, D. pseudocajanus, D. subcapitatus, D. trinervatus, Dorycnium ger-

manicum, Dumasia villosa, Elephantorrhiza elephantina, Elizabetha duckei, Entada abyssinica, E. polyphylla, E. polystachya, E. pursaetha, E. scandens, E. spicata, Enterolobium schomburghkii, E. cyclocarpum, E. timbouva, Eriosema affine, E. burkei, E. chinense, E. cordatum, E. engleranum, E. glabrum, E. grandiflorum, E. heterophyllum, E. longifolium, E. pauciflorum, E. polystachyum, E. salignum, Eriosema spp. (2), *E. violaceum, Eminia antennulifera, Erythrina falcata, E. latissima, E. reticulata, E. rubrinerva, E. polyadenia, Erythrophloem africanum, Eschweilera timbuchensis, Euchresta horsefieldii, Eysenhardtia amorphoides, E. polystachya, Flemingia bracheata, F. congesta, F. grahamiana, F. strobilifera, Galactea apiifolia, G. macrophylla, G. pretiosa, G. speciosa, Gleditsia aquatica, G. dorrida, Gleditsia sp., Gliricidia sepium, Glottidium vesicarium, Glycine javanica, G. koidzumii, G. max, G. soya, G. wightii, Glycyrrhiza lepidota, Gompholobium uncinatum, Guibourtia conjugata, Gustavia brasiliana, G. poeppigiana, Gymnocladus dioicus, Haematoxylon campechianum, H. brasiletto, Hardenbergia retusa, Harpalia brassiliensis, Hedysarum sp., Heterostemon mimosioides, Hoffmannseggia sandersoni, H. densiflora, H. melanosticta, Holocalyx balanse, Hovea longipes, Humularia elisabethvilleana, Hymenaea courbaril, H. strigonocarpa, Hymenaea sp., Hymenolobium excelsum, H. heterocarpum, H. petraeum, H. oblongifolia, Hypocalyptus obcardatus, Indigofera altenans, I. angustifolia, I. antunesiana, I. argyroides, I. arrecta, I. astragalina, I. atriceps, I. brevideus, I. butayei, I. caroliniana, I. circinnata, I. comosa, I. coriacea, I. congesta, I. demissa, I. drepanocarpa, I. endocephala, I. enneaphylla, I. gastigata, I. fanshawei, I. frondosa, I. frutescens, I. garkeana, I. hedyantha, I. heterophylla, I. hewittii, I. hirsuta, I. hilaris, I. holubii, I. homblei, I. inhambanensis, I. linifolia, I. longibarbata, I. lupatana, I. malacostachys, I. melanadenia, I. microphylla, I. microcarpa, I. nebrowniana, I. nummularifolia, I. pongolana, I. podocarpa, I. rhynchocarpa, I. schinzii, I. schimperi, I. sessilifolia, I. setiflora, I. sordida, I. spicata, I. stenophylla, I. stenoptera, I. stricta, I. subfruticosa, I. subulata, I. subulifera, I. swaziensis, I. teita, I. tesmanii, I. tinctoria, I. tomentosa, I. tristis, I. tristoides, I. vicioides, I. welwitchii, I. williamsonii, I. zollingeriana, Inga acuminata, I. cinnamomea, I. flagelliformis, I. hartii,*

I. heterophylla, I. ingoides, I. laurina, I. leptoloba, Inga sp. aff. *longifolia, I. marginata, I. nobilis, I. nuda, I. obtusa, I. radians, I. sessilis, I. setifera, Inga spp.* (7), *I. spuria, I. stenoptera, I. thibaudiana, I. velutina, I. venosa, I. vestita, I. vulpina, Inocarpus edulis, Inocarpus spp.* (2), *Intsia bijuga, Isoberlinia tomentosa, Kennedya prostrata, Kotschya aeschynomenoides, K. caitulifera, K. strobilantha, Krameria citisoides, K. cuspidata, K. glandulosa, K. grayi, K. navae, K. ixina, Krameria spp.* (2), *Kummerowia stipulacea, K. striata, Lamprolobium fruticosum, Lathyrus laetiflorus, L. maritimus, L. pratensis, L. vestitus, Lespedeza capitata, L. cuneata, L. hirta, L. intermedia, L. repens, L. sericea, Lespedeza spp.* (2), *L. violacea, L. virginica, Lessertia pauciflora, Leucaena esculenta, L. glabrata, L. glauca, L. leucocephala, L. macrophylla, Leucaena sp., Listia heterophylla, Lonchocarpus capassa, L. castilloi, L. guilleminianus, L. latifolius, L. leucanthus, L. minimiflorus, L. palmeri, L. punctatus, L. sericeus, Lonchocarpus spp.* (4), *L. subglaucescens, Lotononis bainesii, L. eriantha, L. solitudinis, L. leobordia, Lotus corniculatus, L. grandiflorus, L. haydonii, L. namulensis, L. nevadensis, L. oblongifolius, L. scoparius, Luetzenburgia reitzii, Lupinus alpestris, L. caudatus, L. hillarianum, L. velutinus, Lysiloma acapulcensis, L. divaricata, L. sabicu, L. trigemina, Lysiphyllum cunninghamii, Machaerium aculeatum, M. floribundum, M. inundatum, M. kuhlmannii, M. scleroxylon, Macrolobium acaciaefolium, M. campestre, M. gracile, M. multijugum, M. pendulum, Macrolobium sp., Maniltoa plurijuga, Medicago arabica, M. denticulata, M. lupulina, M. minima, M. polymorpha, M. sativa, Medicago spp.* (2), *Melanoxylon braunia, Melilotus alba, M. indica, M. officinalis, Microlespedeza stricta, Millettia auriculata, M. grandis, M. reticula, M. thonningii, M. dielsiana, M. ovalifolia, M. pulchra, M. speciosa, M. usaramensis, Mimosa acapulcensis, M. acerba, M. aculeaticarpa, M. barretoi, M. bimucranota, M. biuncifera, M. brandegeci, M. calodendron, M. calothamnos, M. coerulea, M. congestifolia, M. extensa, M. floculosa, M. futuracea, M. hamata, M. invisa, M. lacerata, M. latescens, M. leucaenoides, M. micropterus, M. multipima, M. niederleinii, M. pigra, M. polyantha, M. pseudoincana, M. ramosissima, M. rigida, M. scabrella, M. schomburgkii, M. sicyocarpa, Mimosa spp.* (20), *M. zygophylla, Moghania grahamina, M. lineata, Mora excelsa, Mucuna altissima,*

M. argyrophylla, M. bennettii, M. capitata, M. coriacea, M. novo-guineensis, M. monosperma, M. poggei, M. prurita, M. rostata, M. sloanei, Muellera frutescens, Myrospermum frutescens, Neorau-tanenia lugardii, Neorautanenia spp. (2), *Neptunia monosperma, N. glazovii, N. plena, N. prostrata, Newtonia hildebrandtii, Nissolia pringlei, N. wislizenii, Olynea testota, Ophrestia retusa, Ormocar-pum bibracteatum, O. orientale, O. trichocarpum, Ormosia arbo-rea, O. occinea, O. macrophylla, O. monosperma, O. nobilis, O. paraensis, Otoptera burchellii, Ougeinia dalbergioides, Parapi-padenia rigida, Pachyrrhizus erosus, P. tuberosus, Parkia auricula-ta, P. biglandulosa, P. pendula, P. ulei, Parkinsonia africana, Pear-sonia atherstonei, Peltogyne cointei, P. porphyrocardia, Peltogyne sp., Peltophorum africanum, P. brasiliense, P. vogelianum, Penta-clethra macroloba, Periandra coccinea, P. dulcis, P. mediterranea, Periandra sp., Petalostemon feayi, P. gracilis, P. pinnatum, Phaseo-lus adenanthus, P. atropurpureus, P. aureus, P. calcaratus, P. for-mosus, P. lunatus, P. mungo, P. pedicillatus, P. pilosus, P. semiexec-tus, P. speciosus, Phaseolus spp.* (8), *P. sublobatus, P. trinervius, Phylacium bracteosum, Phyllocarpus riedellii, Phyllota phylli-coides, Piliostigma thonningii, Piptadenia gonocantha, P. laxa, P. peregrina, Piptadenia spp.* (2), *Pithecellobium acatlense, P. bre-vifolium, P. calostachys, P. cauliflorum, P. dulce, P. elastichophyl-lum, P. latifolium, P. lucidum, P. lusorium, Pithecellobium spp.* (3), *P. unguis-cati, Plathymenia reticulata, Platylobium triangulare, Platymiscium floribundum, P. trinitalis, P. yucatanum, Pongamia glabra, Poiretia latifolia, Poiretia spp.* (4), *P. tetraphylla, Ponga-mia pinnata, Pongamia sp., Prosopis spicigera, Pseudarthria hook-eri, Pseudocardia zambesiaca, Psoralea aphylla, P. bracteata, P. candicans, P. foliosa, P. fruticans, P. hirta, P. lupinellus, P. nu-dum, P. obtusifolia, P. patens, P. pinnata, P. psoraloides, P. repens, P. restidoides, Psoralea sp., P. verrucosa, P. wilmsii, Pterocarpus amazonicus, P. acapulcensis, P. angolensis, P. brenanii, P. indicus, P. officinalis, P. rohrii, P. rotundifolius, Pterocarpus spp.* (2), *P. vidalianus, Pterodon pubescens, Pterodon sp., Pterolobium stel-latum, Pteroloma triquetrum, Pueraria hirsuta, P. lobata, P. thun-bergiana, P. tonkinensis, Pultanea altissima, P. angustifolia, P. gra-veolens, P. mollis, P. scabra, P. stricta, P. tenuifolia, P. microphylla, P. polifolia, Pycnospora lutescens, P. nervosa, Rafnia amplexicau-*

lis, R. angulata, R. axillaris, R. cuneifolia, R. perfoliata, Rhynchosia albiflora, R. australis, R. caribaea, R. clivorum, R. corylifolia, R. cunninghamii, R. cyanospermum, R. effusa, R. longeramosa, R. memnonia, R. minima, R. nervosa, R. phaseoloides, R. pitcheria, R. pyramidalis, R. resinosa, R. sordida, Rhynchosia spp. (4), Robinia hispida, R. pseudoacacia, Saraca declinata, S. indica, Schizolobium amazonicum, S. parahybum, S. excelsum, Schotia afra, S. brachypetala, S. capitata, Serianthes sp., Sesbania cannabina, S. exaltata, S. exasperata, S. grandiflora, S. grandifolia, S. macrantha, S. macrocarpa, S. microphylla, S. mossambicensis, S. punicea, S. roxburghii, S. sericea, S. sesban, S. speciosa, Sesbania sp., S. tetraptera, Smithia conferta, S. thymobora, Sophora japonica, Sophora sp., Sphaerolobium vimineum, Sphenostylis angustifolia, S. erecta, S. briartii, S. stenocarpa, Stizolobium (= Mucuna) cochinensis, Strongyolodon macrobotrys, Strophostyles helvola, S. umbellata, Stryphnodendron barbatimam, S. floribundum, S. pulcherrima, Stryphnodendron spp. (2), Stylosanthes gracilis, S. guianensis, Stylosanthes sp., S. montevidensis, S. viscosa, S. fruticosa, S. mexicana, Swainsonia burkittii, S. canescens, S. conferta, S. lessertifolia, S. procumbens, S. luteola, S. macrostachya, Swartzia alterna, S. ingaefolia, S. langsdorffia, S. madagascariensis, S. simplex, S. stipulata, S. stipulifera, S. trinitensis, Sylitra biflora, Tachygalia cavipes, Tamarindus indica, Tephrosia sp. aff. rhondantha, T. capensis, T. cordata, T. dasyphylla, T. decora, T. ehrenbergiana, T. grandiflora, T. lepida, T. leucantha, T. littoralis, T. longipes, T. nana, T. nicaraguensis, T. nitens, T. obovata, T. piscatoria, T. polystachya, T. purpurea, T. rosea, T. sphaerosperma, T. villosa, T. virginiana, T. zombensis, Teramnus volubilis, Thermopsis mollis, Torresia acreana, Tounatea simplex, Trifolium africanum, T. agaricum, T. angustifolium, T. incarnatum, T. involucratum, T. monoense, T. repens, T. resupinatum, Trigonella uncata, Tylosema fassoglensis, Umtiza listeriana, Uraria lagopodoides, Vandasia retusa, Vatairea guianensis, V. sericea, Vataireopsis speciosa, Vicia angustifolia, V. benghalensis, V. cracca, V. faba, V. gigantea, V. hirsuta, V. sativa, V. tetrasperma, Vigna decipiens, V. gazensis, V. longiloba, V. lutea, V. luteola, V. marina, V. sinensis, Vigna sp., V. vexillata, V. wilmsii, Voucapoua americana, Wallaceodendron celebicum, Wiborgia sericea, Wisteria brachynotrys, W. japonica, W. sinensis,

Zollernia ilicifolia, Z. paraensis, Zornia capensis, Z. glochidata, Zornia sp.

REFERENCES

Arora, S. K. (Ed.), *Chemistry and Biochemistry of the Legumes,* Edward Arnold, London (1983).
Salatino, A. and O. R. Gottlieb, *Revista Brasiliera de Botanica 4* (1981) p. 83.

LEMNACEAE
6 genera; 30 species

The duckweeds are found in fresh water throughout the world and are used as ornamentals in aquaria.

Alkaloids are not known for the family, and three samples, *Lemna minor, L. valdiviana,* and *Spirodela polyrhiza,* gave negative tests in this study.

LENTIBULARIACEAE
4 genera; 245 species

This is a family of the marsh vegetation of all continents. Some species are grown in aquaria.

Alkaloids have not been recorded for the family. Of nine species tested here, *Utricularia firmula* was positive, and *U. aurea, U. meticulata, U. stellaris, U. transrugosa, U. vulgaris,* and three undetermined *Utricularia* species were negative.

LILIACEAE
294 genera; 4,500 species

The figures given here for the number of genera and species in this family are those of Cronquist, via Mabberley, who places several other lilylike families in this large group. These plants have a wide distribution, especially in warm temperate and tropical regions. In addition to many ornamentals, crop plants such as asparagus and the onions and their relatives make this family of considerable economic importance.

Alkaloids are common in the family. Among 200 species tested in this study, several plants were known to have been alkaloidal from reports in the earlier literature: *Allium christophi* (sometimes placed in the Alliaceae), *Amianthum* (= *Zigadenus*) *muscaetoxicum*, *A. plumosus* (1/3), *Camptorrhiza strumosa*, *Colchicum luteum*, *Dipidax* (= *Oxinotis*) *triquetra* (1/2), *Gloriosa superba* (1/3), *Smilacina racemosa*, *Veratrum viride*.

The following positive species had not been listed previously as alkaloidal: *Albuca kirkii*, *A. melleri*, *Allium cernuum*, *A. karataviense*, *Asparagus asperagioides* (1/2), *A. laricinus*, *A. officinalis* (1/2), *A. pearsoni*, *A. racemosus* (3/4), *A. retrofractus*, *A. rivalis*, *Allium sp.*, *A. suaveolens* (1/2), *A. thunbergianus*, *A. virgatus*, *Bessera elegans*, *Bulbine frutescens*, *B. narcissifolia* (1/2), *Bulbinella hookeri*, *Burchardia sp.*, *Chlorophytum spp.* (2/3), *Drymophila moorei*, *Eriospermum abyssinicum*, *E. galpinii*, *Hemerocallis eelfin*, *Lachenalia pendula*, *Lilium philadelphicum*, *L. superbum* (1/4), *Schelhammera multiflora*, *S. pedunculata* (1/2), *Scilla chinensis*, *S. natalensis*, *Scilla sp.* (1/4), *Stypandra australis*, *Tofieldia calyculata*, *Urginea* (= *Drimia*) *altissima*, *Veratrum californicum*, *V. woodii*, *Wurmbea spicata* (1/2), *Zigadenus brevibracteatus*, *Z. exaltatus*, *Z. fremontii*, *Z. glaberrinus*.

Negative tests were obtained for the following: *Agapanthus inapterus*, *Agapanthus sp.*, *Albuca altissima*, *A. canadensis*, *Albuca sp.*, *Aletris farinosa*, *A. spicata*, *Allium kunthii*, *A. scaposum*, *Allium sp.*, *triquetrum*, *A. vineale*, *Androcymbium capense*, *Antherium chlamydophyllum*, *A. whytei*, *A. manum*, *Asparagus aethiopicus*, *A. angolensis*, *A. capensis*, *A. cochinchinensis*, *A. falcatus*, *A. lucidus*, *A. multiflorus*, *Asparagus sp.*, *A. sprengeri*, *A. striatus*, *A. undulatus*, *Asphodelus fistulosus,* *Astelia nadeaudi*, *A. neocaledonica*, *A. nervosa*, *A. papuana*, *A. solandri*, *Bowiea volubilis*, *Brodiaea lutea*, *B. pulchella,* *Bulbine abyssinica*, *B. aloides*, *B. asphodelioides*, *B. tortifolia*, *Calochortus barbatus*, *C. clavatus*, *C. kennedyi*, *C. macrocarpa*, *C. purpureus*, *Camptorrhiza hyssopifolium*, *Chlorogalum pomeridianeum*, *Chlorophytum comosum*, *C. papillo-*

sum, C. crispum, C. capense, C. kymatodes, Chlorophytum sp.,
Convallaria sp., Cyanella hyacinthoides (sometimes Techophilia-
ceae), *C. lutea, Danae gallae, Dianella congesta, D. ensifolia, D.*
intermedia, D. javanica, D. sandwicensis, D. tasmanica, Dipcadi
glaucum, Dispermum kawakamii, Drimia alta, D. zombensis,
Echeandia macrocarpa, E. paniculata, E. reflexa, Echeandia sp.,
Eriospermum bellendenii, E. cooperi, Eucomis sp., E. undulata,
Fritillaria meleagris, Gloriosa simplex, G. virescens, Heloniopsis
umbellata, Hemerocallis fulva, Herreria sp., Hesperocallis undula-
ta, Iphigenia indica, Kniphofia ensifolia, K. multiflora, Kniphofia
sp., K. splendida, K. uvaria, Lachenalia unifolia, Lilium canadense,
L. martagon, L. washingtonianum, Liriope muscari, L. spicata, Lit-
tonia rigidifolia, Maianthemum canadense, Massonia latifolia, No-
thoscordum bivalve, N. montevidense, Nothosceptrum (= *Kniphofia*)
andondense, Ophiopogon japonicus, Ophiopogon sp., Ornithogalum
ecklonii, O. saundersiae, O. thyrsoides, O. umbellatum, O. zeyheri,
Polygonatum canaliculatum, P. commutatum, P. officinale, P. runcina-
tum, Pseudogaltonia sp., Rhipogonum scandens, Schelhammera pe-
dunculata, Schizobasis angolensis, Schizobasis sp., Scilla lanceaefolia,
S. megaphylla, S. natalensis, S. rigidifolia, Scilla spp. (2), *S. zambesia-*
ca, Semele androgyna, Smilacina formosana, S. thyrsoidea, Stenan-
thium frigidum, Trachyandra ciliata, T. falcata, T. laxa, Tricyrtis for-
mosana, T. formosana var. *glandulosa, Tulbaghia frutans, T. violacea,*
Tulipa micheliana, T. tarda, Urginea dregei, U. epigea, U. multisetosa,
Uvularia sessilifolia, Xeronema moorei, Xerophyllum tenax.

LIMNANTHACEAE
2 genera; 8 species

This is a small family of North American cultivated ornamentals.
Alkaloids are not known except for a recorded positive test for
Limnanthes douglasii; in this study a sample of *Floerkea prosperi-*
nacoides was negative.

LINACEAE
15 genera; 300 species

The family is cosmopolitan but found mostly in the temperate
zones of both hemispheres.

The genus *Linum* yields flax and linseed oil; other species are used as ornamentals.

Not much is known of the chemistry of the family; the presence of alkaloids has been reported in *Hugonia oreogena* and *H. penicillanthemum* and in two species of *Linum*. Of ten samples representing eight species, *Hugonia orientalis* and *Ochtocosmus lemaireanus* were positive; *Durandea jenkinsii*, *D. pallida*, *Hebepetalum humirifolium*, *Linum orizabae*, *L. schiedeanum*, and *Saccoglottis* aff. *guianensis* were negative.

LOASACEAE
15 genera; 260 species

Most representatives of this primarily American family are found in western South America. A few are in Africa and Arabia.

Little chemical work has been done with the family; *Mentzelia decapetala* is known to be alkaloidal. In this survey, *Mentzelia dispersa*, *M. hispida* (1/2), and *M. laevicaulis* as well as *Petalonyx thurberi* (1/2) gave positive results. Eleven other species were negative: *Cevallia sinuata*, *Eucnide cobata*, *Gronovia longiflora*, *Kissenia capensis*, *Loasa rupestris*, *Mentzelia aspera*, *Mentzelia davisoniana*, *M. hirsutissima*, *M. veatchiana*, *M. wrightii*, *Ortiga* (= *Loasa*) *branea*.

LOGANIACEAE
29 genera; 600 species

This is a pantropical and warm temperate family well represented in South America. It is the source of several types of curares and the drug strychnine, which is now primarily used as a rat poison. Some species are cultivated as ornamentals.

Interest in the arrow and dart poisons of South America and Africa has resulted in a considerable knowledge of the alkaloids of this family. Positive tests obtained on the following species were expected based on earlier studies: *Gelsemium elegans*, *G. sempervirens*, *Strychnos angolensis*, *S. erichsonii*, *S. angustiflora*, *S. bancroftiana*, *S. colubrina* (1/3), *S. hemmingsii*, *S. jobertiana* (stem), *S. stuhlmannii*.

Other positive tests were obtained with the following: *Antho-*

cleista nobilis, A. grandiflora, A. schweinfurthii, Buddleja cordata (1/4), *B. davidii* (1/2), *B. lobulata, B. lindlayana, B. saligna* (1/2), *Emorya suaveolens, Fagraea tahitensis, Gomphostigma virgatum, Mitrasacme alsinoides, Neubergia corynocarpa, Nuxia floribunda* (1/2), *Potalia amara* (2/2), *Strychnos cocculoides, S. coelospermum, S. colubrina* (1/3), *S. innocua* (1/2), *S. lucida, S. mitis, S. panurensis, S. pungens* (2/3), *S. rondeletoides, Strychnos spp.* (4).

Negative tests were obtained with the following species: *Anthocleista zambesiaca, Antonia ovata, Buddleja madagascariensis, B. americana, B. artenifolia, B. asiatica, B. brasiliensis, B. elegans, B. humboldtiana, B. marrubiifolia, B. microphylla, B. parviflora, B. perfoliata, B. saligna, B. salviifolia, B. scordioides, B. sessiliflora, B. speciosissima, Buddleja spp.* (2), *B. vetula, Couthovia corynscarpa, C. seemanii, Fagraea berteriana, F. bodenii, F. ceilanica, F. elliptica, F. racemosa, F. salticola, Geniostoma arfakense, G. lingustrifolium, G. rupestre, Geniostoma sp., G. weinlandii, Labordea tinifolia, Mitrasacme elata, M. pygmaea, Neubergia kochii, Nuxia congesta, N. oppositifolia, Polypremum procumbens, Spigelia anthelmia, S. humboldtiana, S. marylandica, S. splendens, Spigelia spp.* (3), *Strychnos brasiliensis, S. brachyata, S. cogens, S. glabra, S. guianensis, S. hirsuta, S. pedunculata, S. pseudoquina, Strychnos spp.* (4), *S. spinosa, S. tepicensis, S. usambarensis.*

Mitrasacme and *Spigelia* have been assigned to a family of their own, Spigeliaceae by some authorities.

LORANTHACEAE
70 genera; 940 species

Some members occur in temperate zones, but this family is primarily tropical. Its economic importance derives largely from the familiar mistletoe (*Viscum album*).

Alkaloids have been noted in some genera but positive tests may be the result of parasitic growth on other plants that contain alkaloids.

Of 83 species tested, the following gave positive results: *Aspidixia* (= *Viscum*) *angulata* (1/2), *Loranthus kraussianus, Phorandendron brachystachyum* (2/3), *P. commutatum, P. robinsonii, Phoradendron spp.* (2/11), *Struthanthus diversifolius, S. microphyllus, Viscum combreticola* (2/2), *V. eucleae, V. verrucosum.*

Negative tests were obtained with the following: *Amyema miquelii*, *A. quandang*, *A. scandens*, *Amylotheca hollrungii*, *Arceuthobium americanum*, *A. cryptopodum*, *A. vaginatum*, *Aspidixia articulata*, *A. liquidambaricola*, *Dendrophthoe falcata*, *Distrianthes molliflora*, *Hyphear owatarii*, *Korthalsella articulata*, *Loranthus brownei*, *L. dregei*, *L. elegans*, *L. eylesii*, *L. longifolius*, *L. minor*, *L. ngamicus*, *L. oleaefolius*, *L. rubromarginatus*, *L. rubroviridis*, *L. scurrulus*, *Loranthus spp.* (3), *L. subcylindricus*, *L. woodii*, *L. zeyheri*, *Lysania exocarpa*, *Phorandendron bolleanum*, *P. californicum*, *P. carneum*, *P. crassifolium*, *P. flavescens*, *P. forestierae*, *R. macrocarpa*, *P. juniperum*, *P. piperiodes*, *P. tequilense*, *P. velutinum*, *P. villosum*, *Phrygilanthus acutifolius*, *P. palmeri*, *Phthirusa pyrifolia*, *Phthirusa sp.*, *Psittacanthus dichrous*, *Psittacanthus sp.*, *Scurrula liquidambaricola*, *S. rhododendricola*, *S. ritozanensis*, *Struthanthus sp.*, *S. staphylinus*, *Viscum articulatum*, *V. multinerve*, *V. nervosum*, *V. orientale*, *V. rotundifolium*, *Viscum sp.*, *V. subserratum*, *V. tricostatum*, *V. verrucosum*.

A separate family, the Viscaceae, has been created (Cronquist) to accommodate *Phoradendron*, *Viscum*, *Korthalsella*, and *Arceuthobium*.

LYCOPODIACEAE
2-5 genera; 100+ species

The family is cosmopolitan except for very arid regions; it is abundant in tropical and subtropical forests. The spores were formerly used medicinally (lycopodium powder).

The family is known to be alkaloidal (Brossi, 1985). Most of the tests performed in this survey were done on bits of material selected from herbarium specimens.

A total of 134 *Lycopodium* species and their varieties were tested. Those previously known to be alkaloid-positive, and likewise found to be so here, included *Lycopodium alopecuroides* (9/10), *L. carolinianum* (2/6), *L. cernuum* (7/32), *L. clavatum* (18/40), *L. contiguum*, *L. densum* (2/3), *L. flabelliforme* (7/9), *L. guidioides*, *L. inundatum* (8/16), *L. laterale* (1/2), *L. obscurum* (7/19), *L. phlegmaria* (3/3), *L. saururus*, *L. selago* (14/17), *L. serratum* (3/3), *L. stichense* var. *nikoense*, *L. tristachyum* (4/8), *L. verticillatum* (2/2), and *L. volubilis* (6/6).

There were, however, several species which were expected to react positively but which did not. This may have been due to the methods used originally in preparing the herbarium species (see page xi). These samples included *L. alpinum*, *L. fastigatum*, *L. fawcettii*, *L. sabinaefolium,* and *L. stichense.*

The following were also positive: *L. adpressum* (1/2), *L. annoticum* (3/3), *L. annotinum* (6/12), var. *acrifolium* (1/4), var. *pungens* (3/6), var. *aqualupianum*, *L. billardieri*, *L. carinatum* (2/2), *L. clavatum* (13/32), var. *integerrinum*, var. *laurentianum*, var. *megastachyon* (1/4), var. *subremotum*, var. *variegatum*, *L. conplanatum* (7/15), subsp. *anceps* (1/2), var. *novoguinense*, var. *validum*, *L. crassum*, *L. cryptomerianum*, *L. dichotomum*, *L. dielsii*, *L. empetrifolium*, *L. filiforme*, *L. foliosum*, *L. fordii* (2/2), *L. hamiltonii*, *L. hedermannii*, *L. hippuris*, *L. hookeri* (2/2), *L. inundatum* var. *biglovei* (3/6), *L. japonicum*, *L. laterale* (1/2), *L. lereticaule*, *L. linifolium* (1/5), *L. lucidulum* (8/9), *L. mandio-caruera*, *L. multispicatum* (1/2), *L. oltsmannii*, *L. parksii*, *L. passerinoides*, *L. phyllanthum* (4/5), *L. pinifolium* (2/2), *L. pithyoides*, *L. polytrichoides*, *L. porophilum* (2/2), *L. prostratum* (4/5), *L. pseudophlegmaria*, *L. reflexum* (2/3), *L. salvinoides*, *L. samoanum*, *L. scariosum* (1/3), *L. selago* var. *apressum* (2/2), var. *miyostrianum* (2/3), var. *patens* (1/2), var. *recurvum*, *L. sieboldii*, *L. somae*, *Lycopodium* spp. (15/18), *L. squarrosum* (1/4), *L. sitchense* var. *nikoense*, *L. taxifolium* (2/2), *L. tomentosum*, *L. tubulosum*, *L. varium*, *L. veitchii*, *L. warneckei*, *L. wightianum* (2/2).

Several species were negative, possibly due in part to the methods employed in preservation of the specimens: *L. affine*, *L. annotinum* var. *alpestre*, *L. brevibracteatum*, *L. cancellatum*, *L. casaurinoides*, *L. cernuum*, f. *pungens*, var. *salakense*, f. *vulcanianum*, *L. chamaecyparissus*, *L. clavatum* var. *lagopus*, var. *monostachya*, *L. complantam* var. *platyrhizoma*, f. *intermedium*, x. *tristachyum*, *L. funiforme*, *L. hippurideum*, *L. platyrhizoma*, *L. robustum*, *L. sikkimense*, *L. skutchii*, *L. subfalciforme*, *L. trifoliatum*, *L. tuerkheimii*, *L. underwoodianum*, *L. watsonianum*.

REFERENCE

Brossi, A. (Ed.) *The Alkaloids*. 26 (1985) 241, Academic Press, Inc. Orlando, FL.

LYTHRACEAE
26 genera; 580 species

The family is abundantly represented in the American tropics but has wide distribution elsewhere, including some species in temperate zones. In addition to use as ornamentals, one genus, *Lawsonia*, has been the source of the dye henna since ancient times.

Little of the chemistry of the Lythraceae was known prior to the 1960s beyond reports of the qualitative presence of alkaloids in a few genera. With the recognition of the novel alkaloid structures in *Decodon* and *Heimia*, an examination of material from other genera gleaned from herbarium specimens was undertaken. The published results are included in those reported here.

Two hundred and thirty samples tested included several known from earlier literature to be positive: *Decodon verticillatus* (17/17), *Heimia myrtifolia* (18/18), *H. salicifolia* (10/13), *Legerstroemia indica* (1/3), *L. speciosa* (1/3), *Lawsonia lanceolatum* (2/3).

The following were also positive: *Antherylium* (= *Ginoria*) *rohrii* (2/2), *Cuphea lanceolata* (1/7), *Cuphea spp.* (3/21), *Ginoria ginoroides*, *G. glabra* (2/3), *G. nudiflora* (2/6), *Haitia buchii*, *H. pulchra*, *Lawsonia inermis* (1/7), *Lythrum acinifolium* (1/2), *L. hyssopifolia* (3/3), *Lythrum spp.* (2/2), *L. lineare*, *L. maritimum*, *Nesaea longipes*, *Rotala ramosior* (1/3), *Woodfordia fruticosa* (1/4).

These members of the family were negative: *Adenaria floribunda*, *Alzatea verticillata*, *Ammania auriculata*, *A. baccifera*, *A. coccinea*, *A. multiflora*, *A. prieriana*, *A. ramosoir*, *A. senegalensis*, *A. teres*, *Crenea surinamensis*, *Cuphea aequipetala*, *Cuphea sp.* aff. *jorullensis*, *C. bustamantha*, *C. carthagenesis*, *C. cyanea*, *C. hookeriana*, *C. hyssopifolia*, *C. ingrata*, *C. itzocamensis*, *C. jorullensis*, *C. linarioides*, *C. linifolia*, *C. llavea*, *C. lobophora*, *C. lutea*, *C. maculata*, *C. mesostemon*, *C. petiolata*, *C. philombria*, *C. pinetorum*, *C. procumbens*, *C. racemosa*, *C. salicifolia*, *C. speciosa*, *C. stygialis*, *C. wrightii*, *Diplusodon buxifolius*, *D. candollei*, *D. crulsianus*, *Diplusodon spp.* (6), *Galpinia transvaalica*, *Ginoria americana*, *G. koehneana*, *G. spinosa*, *Grislea* (= *Pehria*) *secunda*, *Lafoensia nummularifolia*, *L. pacari*, *L. punicifolia*, *Lafoensia sp.*, *L. speciosa*, *Lagerstroemia parviflora*, *L. pyriformis*, *Lagerstroemia spp.* (2/2), *L. speciosa*, *Lawsonia alba*, *Lythrum californicum*,

L. flexuosum, L. gracile, L. salicaria, Nesaea heptamera, N. hispidula, N. schinzii, Orias excelsa, Pehria compacta, Pemphis acidula, Peplis glabra, Peplis spp. (2), *Physocalymna sp., Pleurophora pungens, Rotala indica, Rotala sp., R. stagiana.*

M

MAGNOLIACEAE
7 genera; 200 species

A tropical to warm temperate family with some extension into north temperate zones (e.g., the tulip tree, *Liriodendron tulipifera*); the magnolias are familiar in our gardens and city parks. Some from this family are used as timber.

The family is known to be alkaloidal: *Liriodendron tulipifera* (4/4), *Magnolia acuminata, M. grandiflora* (6/10), *M. kobus,* and *Michelia champaca* were positive as literature references had indicated. *Elmerilla papuana, Magnolia coco, M. fraseri, M. pyramidata, M. sieboldii* (2/2), *Magnolia spp.* (3/3), *M. stellata* (3/3), *M. virginiana* (4/4), *M. forbesii, M. fuscata* (2/3), and *Talauma mexicana* (1/4) were also positive.

There were, however, negative results as follows: *Illicium arborescens, I. floridanus, I. leucanthum, I. parviflorum, Illicium sp., I. verum* (the genus is considered as a separate family, Illiciaceae, by some taxonomists), *Kadsura japonica* (in Schizandraceae by some authorities), *Magnolia tripetala, M. obovata,* and *Michelia compressa formosana.*

MALPIGHIACEAE
68 genera; 1,100 species

This is a family of the warm regions, especially those of South America, where *Banisteriopsis* and its relatives supply one of the major hallucinogenic preparations of the western Amazon area.

Alkaloids, particularly ∝-carbolines, are found in the family though not so frequently as one might expect. Tests on 114 samples representing 80 species gave the following positives: *Banisteriopsis caapi* (1/2), *Cabi* (= *Callaeum*) *paraensis* (both previously known); *Banisteria* (= *Heteropteris*) *campestris* (1/2), *B. cotinifolia, Bunchosia glandulifera, B. lindeniana, B. palmeri* (2/3), *Dicella bracteosa* (1/2), *Echinopteris glandulosa* (1/4), *Malpighia glabra* (1/3), *Mascagnia macroptera, M. ovata.*

Negative tests were obtained with the following: *Acridocarpus natalitus, Aspidocarpa sericea, Banisteria* (= *Heteropteris*) *campestris, B. cotinifolia, B. muricata, B. palmeri, Banisteria spp.* (2), *Banisteriopsis lucida, Brachypteris ovata, Bunchosia nitida, B. prismatocarpa, Burdachia spaerocarpa, Byrsonima amazonica, B. ciliata, B. coriacea, B. crassa, B. crassifolia, B. cuneata, B. japurensis, B. laxiflora, Byrsonima spp.* (4), *B. spicata, B. verbascifolia, Camarea affinis, Camarea sp., Diacidia parviflora, Dicella bracteosa, Diplopteris sp., Galphimia glauca, Gaudichaudia filipendula, Heteropteris aceroides, H. aenea, H. macrostachya, H. martiana, Hiraea dipholiphylla, H. velutina, Malpighia sp., Mascagnia concinna, M. ovatifolia, M. polybotrya, Mascagnia sp., Peixoto tomentosa, Rhyssopteris sp., R. timoriensis, Sphedamnocarpus angolensis, S. galphimifolius, S. pruriens, Stigmaphyllon sp.* cf. *brachiatum, S. ciliatum, S. convolvulatum, Stigmaphyllon sp.* aff. *fulgens, S. littorale, Tetrapteris discolor, T. salicifolia, Tetrapteris sp., T. squarrosa, Thryallis dasycarpa, T. glanea, T. glauca, Triaspis leedertziae, T. macropteron, T. rogersii, Tristellateia australasiae, Tristellateia sp.*

REFERENCE

Schultes, R.E. and R.F. Raffauf, *The Healing Forest*, Dioscorides Press, Portland, OR (1990).

MALVACEAE
116 genera; 1,550 species

This is a cosmopolitan, but especially tropical, family. It is known as the source of cotton and many species of hibiscus used as ornamentals and the makings of a drink common in eastern Africa. The vegetable okra, *Abelmoschus esculentus*, also comes from this family.

Alkaloids are not common; of the 218 species tested in this study, two had been reported to contain them: *Sida cordifolia* (1/7) and *Urena lobata* (1/10). Others that gave positive tests included: *Abutilon angulatum, A. hirtum, A. indicum* (1/2), *A. ramosum, Hibiscus astromarginatus, H. elliotiae, H. manihot* (1/2), *H. mastersianus* (1/2), *H. surrathensis, Malva pusilla, Malvastrum lacteum, Pavonia cancellata, Sida chrysantha, S. serratifolia, Sphaeralacea angustifolia* (2/5), *S. hastulata* (1/3), *Thespesia lampas* (1/2), *T. populnea* (1/6).

Most tests were negative: *Abelmoschus moschatus, Abutilon andrewsianum, A. austro-africanum, A. calliphyllum, A. crispum, A. ellipticum, A. englerianum, A. giganteum, A. grandiflorum, A. helmsleyanum, A. hypoleucum, A. incanus, A. leucopetalum, A. pauciflorus, A. pictum, A. polyandrum, A. pycnodon, A. ramiflorum, A. sonneratianum, A. striatum, Anoda hastata, A. incarnata, A. parviflora, Anoda sp., Azanza garckeana, Bakeridesia macrocarpa, B. rufinerva, Bastardia viscosa, Bastardiopsis densiflora, Bogenbardia crispa, Cienfuegosia digitata, C. hildebrandtii, C. gossypifolia, C. heterophylla, Gaya grandiflora, Gossypium barbadense, G. herbaceum, G. hirsutum, G. purpureum, Hampea appendiculata* var. *longicalyx, H. euryphylla, H. integerrima, H. latifolia, H. micrantha, H. nutricia, H. punctulata, H. rovirosae, H. sphalocarpa, H. stipitata, H. tomentosa, H. thespesioides, H. trilobata, Hibiscus aculeatus, H. allenii, H. arnottianus, H. articulatus, H. attenuatus, H. bifurcatus, H. caesius, H. calyphyllus, H. cannabinus, H. cardiophyllus, H. denudatus, H. diversifolius, H. dongolensis, H. engleri, H. esculentus, H. ficulneus, H. fuscus, H. hierianus, H. hoodlandianus, H. irritans, H. meeusei, H. moscheutos, H. mutabilis, H. panduriformis, H. prateritus, H. rhondanthus, H. rosasinensis, H. sabdariffa, H. schizopetalus, H. schinzii, H. sidiformis, H. sinensis, Hibiscus* spp. (9), *H. spiralis, H. sturtii, H. syriacus, H. tiliaceus, H. trionum, H. tubiflorus, H. vitifolius, Hoheria populnea, Iliamna bakeri, Kosteletzyka buettneri, Kydia calycnia, Lagunaria patersonii, Lavatera cretica, L. kashmiriana, L. plebeja, Malachra alceifolia, M. capitata, M. fasciata, Malacothamnus fasciculatulus, Malissoa corymbosa, Malva neglecta, M. parviflora, Malva sp., Malvastrum rotundiflora, M. spicatum, M. coromandelianum, Malvaviscus arboreus, M. conzatii, M. grandiflorus, Malveopsis grossulariaefolium, M. scabrosa,*

Malveopsis sp., *Modiola caroliniana*, *Modaliastrum malvifolium*, *Montieroa ptarmicifolia*, *Notoxylinon australe*, *Pariti* (= *Hibiscus*) *tiliaceum*, *Pavonia columella*, *P. erythrolema*, *P. fruticosa*, *P. garck- eana*, *P. hirsuta*, *P. leptocalyx*, *P. malvacea*, *P. melanommata*, *P. microphylla*, *P. missionum*, *P. paniculata*, *P. patens*, *P. polymor- pha*, *P. praemorsa*, *P. schrankii*, *Pavonia spp.* (5), *P. transvaalensis*, *Peltaea acutifolia*, *Pseudabutilon spicatum*, *Radyera urens*, *Robinso- nella mirandai*, *Sida acuta*, *S. cordifolia*, *S. carpinifolia*, *S. corruga- ta*, *Sida sp.* aff. *corrugata*, *S. dregei*, *S. filipes*, *S. glutinosa*, *S. her- maphrodita*, *S. hoepfnerii*, *S. humilis*, *S. linifolia*, *S. macrodon*, *S. multicaulis*, *S. palmeri*, *S. paniculata*, *S. procumbens*, *S. rhombifo- lia*, *Sida spp.* (3), *S. spinosa*, *S. subspicata*, *Sidalacea multifida*, *S. oregana*, *Sphaeralacea ambigua*, *S. angustifolia*, *S. elegans*, *S. orcuttii*, *S. parviflora*, *S. subhastata*, *S. vitifolia*, *Thespesia acutilo- ba*, *T. patellifer*, *Triclisia trifolia*, *Typhalaea fruticosa*, *Wissadula amplissima*, *W. contrata*, *W. paraguayensis*.

MARANTACEAE
31 genera; 550 species

The family is well distributed in the American tropics. It is the source of arrowroot, a wax, and fibers used to weave baskets. There has been recent interest in *Thaumatococcus* as a source of a protein 3,000 times sweeter than sucrose.

Little is known of the chemistry of the family. Seventeen samples including 14 species were tested to give two positive results: *Calathea allouia*, *Thalia geniculata*. The remainder were negative: *Calathea cyclophora*, *C. insignis*, *C. macrosepala*, *Donax cannaeformis*, *Ischno- siphon obliquus*, *Maranta arundinacea*, *M. gibba*, *Maranta sp.*, *My- rosma connifolia*, *Phrynium sp.*, *Stromanthe tonckat*, *Thalia sp.*

MARCGRAVIACEAE
5 genera; 108 species

This is a family of tropical America; little of its chemistry is known. Eight samples and eight species were tested for alkaloids with negative results: *Marcgravia hartii*, *M. elegans*, *M. norantes*,

M. tobagensis, Norantea brasiliensis, Souroubea exauriculata, S. guianensis, S. pachyphylla.

MARTYNIACEAE
18 genera; 95 species

Many taxonomic assignments in this New World tropical family are questionable. Cronquist has the family as Pedaliaceae with one species of *Martynia* while others have two. No positive results were obtained on testing samples labeled *Martynia annua* and *Proboscidea fragrans.*

MAYACACEAE
1 genus; 4 species

These are South American fresh water aquatics, with one species in Angola. Alkaloids are not known nor were they found in one undetermined species of *Mayaca.*

MELASTOMATACEAE
215 genera; 4,750 species

This is a family of tropical and warm regions, especially of South America. It yields lumber, dyes, and some ornamentals.

Alkaloids are not known in the family; only three of 194 species tested in this survey gave positive results: *Cambessedesia sp., Clidemia rubra* (1/2), and *Tibouchina longifolia* (1/2). The remainder were negative: *Acanthella sprucei, Aciotis acutiflora, Acisanthera sp., Adelbotrys marginata, Arthrostema fragile, A. macrodesmum, Arthrostema spp.* (2), *Astronia sp., Bellucia grossularioides, B. imperialis, Blakea rosea, Cambessedesia hilariana, Centradenia salicifolia, Clidemia capitellata, C. hirta, C. involucrata, Clidemia sp.* cf. *sphanatha, Conostegia icosandra, C. xalepensis, Dissotis canescens, D. debilis, D. multiflora, D. phaeotricha, D. princeps, D. pulchra, Dissotis sp., Graffenrieda caryophyllacea, G. rupestris,*

*Graffenrieda sp., Henriettea multiflora, H. ramiflora, Henriettea
sp., Heterocentron alatum, H. occidentalis, Lavoisiera caryophylla,
L. cordata, L. gentianoides, L. phyllodycina, L. pulchella, Lavoisiera spp.* (3), *Leandra australis, L. barbinervis, L. echinata,
L. melanodesma, L. pectinata, L. polystachya, L. rufescens, Lean-
dra spp.* (4), *Loreya acutifolia, Macairea schultesii, Maieta guia-
nensis, M. malabaricum, M. polyanthum, M. sanguineum, M. sep-
temnervium, Melastoma sp., Memecylon edule, M. tinctorium,
M. umbellatum, Meriania sp., Miconia acinodendron, M. albicans,
M. amazonica, M. aplostachya, M. chrysophylla, M. cilianta,
M. cinerascens, M. cinnamomifolia, M. chrysophylla, M. discolor,
M. disparitis, M. dodecandra, M. erythronitha, M. guianensis,
M. hyemalis, M. kleinii, M. klugii, M. lucera, M. laevigata,
M. lanata, M. lepidota, M. lingustrina, M. lingustroides, M. macro-
tis, M. magnifica, M. malagriphylla, M. mexicana, M. minutiflora,
M. mucronata, M. myriantha, M. nervosa, M. paulensis, M. pluke-
netii, M. poeppigii, M. sellowiana, M. serratula, Miconia spp.* (12),
*M. stenostachya, M. theaezans, M. tomentosa, M. tristis, M. wittii,
Microlepis majuscula, M. oleaefolia, Microlicia spp.* (3), *Mono-
chaetum calcaratum, M. depennum, M. pringlei, Mourira chamis-
soniana, M. grandiflora, M. huberi, M. myrtifolia, M. segotiana,
Mourira sp., Myriospora egensis, Nepsera aquatica, Osbeckia chi-
nensis, O. crinata, Ossaea amygdaloides, Pachycentria formosana,
oxyspora paniculata, Pachyloma coriaceum, Pterolepis glomerata,
Pterolepis sp., Rhexia mariana, R. virginica, Tetrazygia bicolor,
Tibouchina aeopogon, T. bourpacana, T. cerastifolia, T. clavata,
T. dubia, T. hatschbachii, T. holosericea, T. kleinii, T. lepidota,
T. martiusiana, T. mexicana, T. monticola, T. naudinienna, T. pilosa,
T. purpussii, T. reitzii, T. schiediana, T. sellowiana, T. sellowii,
Tibouchina spp.* (12), *T. ursina, Tococa aristata, Trembleya parvi-
flora, Tristemma incompletum, T. virusanum.*

MELIACEAE
51 genera; 575 species

A tropical family with a few subtropical representatives, the Me-
liaceae yield lumber, including mahogany, and some ornamentals.

Alkaloids have been reported in several genera but no particular type appears to be characteristic. In the study reported here, two species reported earlier to contain alkaloids were found positive: *Entandophragma caudatum* (4/5) and *Melia azadirach* (6/23). Other positive species included the following: *Aglaia goebeliana, Amoora rohituka, Azadirachta indica* (2/3), *Cipadessa fruticosa, Dysoxylum binectariferum, D. rufum, Ekebergia capensis* (1/2), *E. meyeri, E. pterophylla, Guarea rusbyi* (1/3), *G. trichiloides* (2/2), *Pteroxylon obliquum* (2/2) (assigned by some authorities to a small South African family, Pteroxylaceae), *Trichilia havanensis, T. hirta, T. pallida, T. prieuriana, T. roka.*

Negative results were obtained with the following: *Aglaia elliptifolia, A. odorata, A. samoensis, A. sepindina, Aglaia sp., Carapa guianensis, Cedrela mexicana, C. odorata, C. sinensis, Cedrela sp., Chukrasia tabularis, Cipadessa baccifera, Duvalia radiata, Dysoxylum arborescens, D. caulostachyum, D. gaudichaudianum, D. lenticillare, D. oppositifolium, D. pachyphyllum, D. pettigrewianum, D. spectabile, Dysoxylum spp.* (6), *D. variable, Ekebergia beneguelensis, Entandophragma sp., Guarea excelsa, G. glabra, G. guara, G. guidonia, G. tuerckheimia, G. verrucculosa, Khaya nyasica, Lansium domesticum, Melia sp., Nymania capensis, Swietenia humilis, S. macrophylla, S. mahagoni, Swietenia sp., Toona ciliata, Trichilia casarettii, T. elegans, T. emetica, T. minutiflora, T. parvifolia, T. raraimana, Trichilia spp.* (4), *T. stellatomentosa, T. trinitensis, T. triphyllaria, Turraea floribunda, T. nilotica, T. oblancifolia, T. obtusifolia, Vavaea spp.* (2), *Xylocarpus granatum.*

MELIANTHACEAE
2 genera; 8 species

These two genera of tropical and southern Africa furnish decorative shrubs and trees. *Greyia,* in this family by Hegnauer and in a family of its own (Greyiaceae) by another authority, has been found to be alkaloidal but two species of *Bersama* (*B. transvaalensis* and *B. tysoniana*) and four of *Melianthus* (*M. comosus, M. insignis, M. major,* and *M. villosus major*), along with *Greyia radlkoferi,* were negative.

MENDONCIACEAE
2 genera; 80 species

No alkaloids have been found in this small family of tropical America and Africa. Four species of *Mendoncia* tested in this study were alkaloid-negative: *M. coccinea*, *M. hoffmannseggiana*, *M. sellowiana*, *Mendonica sp.*

MENISPERMACEAE
78 genera; 570 species

A few species of this alkaloidal family of tropical and warm climates extend into temperate zones. Various genera/species have been used as fish poisons, medicinals, contraceptives, sweeteners, and a few ornamentals.

The alkaloids of several genera have been identified and their plant sources included in this study were also positive: *Abuta grandiflora* (3/3), *Antizoma angustifolia*, *Chondrodendron toxicoferum* (2/2), *C. tomentosum*, *Cissampelos mucronata* (4/4), *C. pareira* (4/6), *Cocculus hirsutus*, *Fawcettia tinosporoides*, *Menispermum canadense*, *Pachygone pubescens*, *Pycnarrhena ozantha*, *Sarcopetalum harveyanum*, *Stephania abyssinica*, *S. hernandifolia*, *S. japonica*.

Other species were likewise positive: *Abuta sp.*, *Antizoma capensis*, *Cissampelos ovalifolia*, *Cissampelos spp.* (2/2), *C. torulosa*, *Cocculus laurifolius* (3/3), *Hyperbaena mexicana*, *Legnephora moorei*, *Pleogyne cunninghamii*, *Pycnarrhena australiana*, *Pycnarrhena sp.* aff. *lucida*, *Sphenocentrum jollyanum* (stem, bark, wood), *Stephania glabra* (3/3), *Stephania sp.*, *Tiliacora warneckei*, *Tinospora caffra*, *T. gragosum* (1/2), *T. sinensis*.

The following species were negative: *Abuta griesbachii*, *Anomospermum folivianum*, *A. schomburgkii*, *Chondrodendron platyphyllum*, *Cissampelos andromorpha*, *Cocculus tribulus*, *Parabaene tuberculata*, *Stephania sarmentosa*, *Tinospora cordifolia*, *T. smilacina*, *Zanania indica*.

MENYANTHACEAE
5 genera; 40 species

This is an aquatic family furnishing a few local medicinals, some cultivated ornamentals, and a number of weeds. Alkaloids are not known in the family; an unidentified species of *Nymphoides* (= *Limnanthemum*) gave a positive test in this study.

MONIMIACEAE
35 genera; 450 species

The Monimiaceae comprise an alkaloidal family of warm and tropical regions that furnishes some ornamentals and timber used mostly locally. They have little economic importance.

Several collections tested in this study were known to contain alkaloids: *Daphnandra dielsii* (2/2), *D. tenuipes, Doryphora aromatica, D. sassafras, Dryadodaphne novoguinense, Laurelia novozealandiae* (8/8), *Palmeria arfakiana* (2/2), *P. fengeriana, Siparuna guianensis* (4/5).

Other positive species included *Kibara papuana, Kibara sp., Levieria spp.* (3/3), *L. acuminata* (1/2), *Siparuna amazonica, S. ternata* (bark) (2/2), *Xymalos monospora*.

Negative tests were obtained with the following: *Hedycarya arborea, Levieria nitens, Mollinedia laurina, Mollinedia sp., Siparuna sp.* aff. *nicaraguensis, Siparuna spp.* (5), *Siparuna* cf. *thecaphora, Trimenia papuana*.

MORACEAE
48 genera; 1,200 species

A family of tropical and warm areas with some in temperate regions, the Moraceae furnish several edible fruits, including mulberries and figs. The well-known and economically important hops as well as *Cannabis* were previously included in this family.

Alkaloids and alkaloid-like substances had been reported and were found in a few of the species tested here, including *Cannabis sativa, Ficus carica, F. hispida, F. septica, Maclura pomifera* (1/4), and *Morus alba* (1/8).

Other alkaloid-positive species included *Brosimum elicastrum, Brosimum sp.* (1/3), *Cardigoyne* (= *Maclura*) *africana, Dorstenia psilurus, Ficus anthelmintica, Ficus sp.* aff. *paraensis, F. copiosa, F. cordata* (1/2), *F. irritans, Ficus spp.* (1/19), *F. sterocarpa, Helicostylis tomentosa* (1/4), *Olmediophaena* (= *Maquira*) *sp., Sorocea klotychiana.*

Cannabis, along with *Humulus,* are now considered members of a separate family, Cannabidaceae.

Negative tests were obtained with the following species: *Artocarpus altilis, A. communis, A. integrifolius, A. lignanensis, A. vriesianus, Bagassa guianensis, Batocarpus amazonicus, Brosimopsis lactescens, Brosimum ducke, B. gaudichaudii, B. velutinum, Brousonettia kaempferi, B. papyrifera, B. kazinoki, Castilla elastica, Celtis africana, Chlorophora spp.* (2), *C. tinctoria, Clarisia spp.* (2), *C. ilicifolia, C. racemosa, C. nitida, Cudrania javanensis, Dorstenia asaroidea, D. contrayerba, D. hispida, Dorstenia spp.* (2), *Ficus asperrima, F. aurea, F. beecheyana, F. bellengeri, F. benjamina, F. broadwayi, F. burkei, F. burtt-davyi, F. capensis, F. capraefolia, F. citrifolia, F. columnaris, F. congesta, F. cotinifolia, F. craterostoma, F. cumingii, F. cunia, F. dammaropsis, F. elastica, F. enormis, F. erecta, F. formosana, F. gameleira, F. gibbosa, F. glomerata, F. glycicarpha, F. gnaphalocarpa, F. godeffroyi, F. hippopotomi, F. heterophylla, F. hirta, F. ierensis, F. ingens, F. involuta, F. isiala, F. laevigata, F. mathewsii, F. kusanoi, F. maxima, F. microchlamys, F. natalensis, F. nekbuda, F. nervosa, F. nitida, F. obliqua, F. padifolia, Ficus sp.* aff. *paraensis, F. pertusa, F. petersii, F. prolixa, F. pumila, F. pungens, F. pyriformis, F. quercifolia, F. radula, F. religiosa, F. retusa, F. rumphii, F. sansibarica, F. soldanella, F. sonderi, F. subnervosa, F. stuhlmanni, F. sycamorus, F. tecolutensis, F. terminalis, F. tjaela, F. tobagensis, F. vaccinoides, F. variegata, F. vasculosa, F. verruculosa, F. vitiensis, F. wightiana, Helianthostylis sp., Helicostylis pedunculata, Helicostylis sp.* aff. *pedunculata, Humulus japonicus, Maclura cochinchinensis, Malaisia scandens, Morus lactea, M. nigra, Paratrophis australiana, Perebea laurifolia, Pseudolmedia oxyphyllaria, Pseu-*

dolmedia sp., Sorocea bonplandii, S. ilicifolia, Sorocea sp., Streblus asper, Streblus glaber, S. urophyllus, Trophis racemosa.

MORINGACEAE
1 genus; 14 species

The genus is found in the semiarid regions of Asia and Africa. *Moringa oleifera* is cultivated in Florida as the "horseradish tree." Glucosinolates and amino acids are found in the family but no alkaloids in the strict sense. Positive tests had been recorded earlier for *Moringa oleifera* and *M. pterygosperma* but retesting of these species along with a sample of *M. concanensis* gave negative results.

MUSACEAE
2 genera; 42 species

The family is native to the Old World tropics and, along with useful fibers, has given us the banana in all of its many varieties. Recent revisions have split the family into two additional families, Heliconiaceae and Strelitziaceae, but these are maintained here in the Musaceae.

Serotonin, a tryptamine, and related compounds have been found in the skins of bananas.

In this study, a positive test for alkaloids was given by *Strelitzia parvifolia*; all other plants tested were negative: *Heliconica hirsuta, H. latispatha, H. psittacorium, Heliconia spp.* (3), *Musa sapientum, Ravenala madagascariensis, Strelitzia reginae, S. nicolai.*

MYOPORACEAE
5 genera; 220 species

This family could be described as a Pacific family with most representatives in Australia and the Indian Ocean. A few species occur in the West Indies. They are used as ornamentals.

Traces of alkaloids had been reported in species of *Myoporum*. One of these, *M. sandwicense*, was likewise positive in the tests reported here along with *M. crassifolium, Myoporum sp.*, and *Oftia africana*.

These species were negative: *Eremophila bignoniiflora, E. freelingii, E. gilesii, E. latrobei, E. strehlowii, Myoporum acuminatum, M. deserti, M. insulare.*

MYRICACEAE
3 genera; 50 species

The family is widespread, almost cosmopolitan. The wax of the fruit of some members is used to make bayberry candles, and many fruits of the family are eaten.

Traditional alkaloids have not been found, but spermidine amides have been isolated from the pollen of *Myrica gale*. In the tests reported here, *Myrica asplenifolia* gave a positive test; the following did not: *Comptonia peregrina, Myrica burmannii, M. californica, M. cerifera, M. conifera, M. cordifolia, M. inodora, M. mexicana, M. oblongata, M. pilulifera, M. rubra, M. serrata, M. tomentosa.*

MYRISTICACEAE
19 genera; 440 species

From lowland rain forests come nutmeg and mace; some members of the family yield a wax of limited use and a hallucinogen of the northwest Amazon.

Members of the family contain many substances of possible interest as bioactive compounds, but the tryptamines of species of Amazonian *Virola* have had special attention for their hallucinogenic effects.

At the time this alkaloid survey was undertaken, several members of the family now known to contain alkaloids had not been studied. Positive tests on these are included in the results reported here. The following species were alkaloid-positive: *Gymnacranthera paniculata, Knema communis* (1/2), *Myristica cagavenensis*,

Virola calophylla, V. callophylloidea, Virola spp. (3/4), *V. surinamensis*.
The following species were negative: *Compsoneura racemosa, Dialyanthera* (= *Otoba*) *parvifolia, Irianthera paraensis, Irianthera sp., Knema conferta, K. hookeriana, K. intermedia, K. latericia, Myristica castanaefolia, M. fragrans, M. grandifolia, M. surinamensis, M. wallichii, Pycnantha schweinfurthii, Virola carinata, V. odorifera, V. oleifera, V. sebifera*.

MYROTHAMNACEAE
1 genus; 2 species

This is a unigeneric family of South Africa and Madagascar. Alkaloids are not known; *Myrothamnus flaballifolia* was negative when tested in this study.

MYRSINACEAE
37 genera; 1,250 species

Some members of this tropical to warm temperate family have had use as ornamentals. Some occur in the Old World, but most are New World species.

Positive alkaloid tests have been recorded for *Maesa* and *Rapanea*, but amino acids are most characteristic of the family.

A few positive results were obtained on testing of 92 samples representing 56 species: *Ardisia crispa* (1/2), *Maesa japonica* (1/3), *M. lanceolata* (1/4), *Myrisine lessertiana, Tapeinosperma sp.*

The following were negative: *Aegiceras corniculatum, Ardisia compressa, A. carnata, A. cornudentata, A. crenata, A. escallinioides, A. japonica, A. liebmannii, A. orenata, A. punctata, A. quinquegona, Ardisia spp.* (3), *A. squamulosa, Conomorpha* (= *Cybianthus*) *peruviana, Cybianthus spp.* (3), *Embelia australiana, E. laeta, E. oblongifolia, Embelia spp.* (2), *Heberdenia penduliflora, Maesa formosana, M. neocaledonica, M. perlarius, M. tenera, Myrsine africana, M. australis, M. divaricatus, M. lessertiana, M. salicina, M. seguine, Parathesis melanosticta, P. serrulata, Pororoca sp.*,

Rapanea ferruginea, R. guianensis, R. jurgensenii, R. melanoph-loeos, R. neriifolia, Rapanea spp. (4), *R. umbellata, R. vaccinioides, Tapeinosperma sp., Weigeltia surinamensis.*

MYRTACEAE
120 genera; 3,850 species

A warm-temperate family, most strongly represented in Australia, the Myrtaceae have given us *Eucalyptus*, edible fruits, spices, medicinals, and ornamentals.

Positive alkaloid tests have been reported for several genera but no alkaloid has yet been isolated and characterized. A few positive tests were obtained in this study: *Eucalyptus globosus, Eugenia anthera, E. atropunctata, E. crenulata* (1/2), *E. dominguensis, Myrcia breviramis, M. citrifolia, M. fallax* (1/2), *Pimenta brava, Pseudocorynophyllus acuminatus, Syzygium malaccensis.*

Most of the 300-plus samples tested were negative: *Acmena* (= *Syzygium*) *acuminatissima, A. smithii, Angophora costata, Baeckea frutescens, B. ramosissima, Baeckea sp., Britoa* (= *Campomanesia*) *acida, Callistemon brachyandrus, C. citrinus, C. lanceolatus, C. phoeniceus, C. vineralis, Calycopus glabra, Calyptranthes concinna, C. paniculata, C. reitzii, Calyptranthes sp., Calyptrogenia hatschbachii, C. microphylla, Calythrix tetragonia, Campomanesia aurea, C. hatschbachii, C. aromatica, C. caerulescens, Cleistocalyx operculatus, Darwinia micropetala, Decaspermum forbesii, D. fruticosum, Eucalyptus bridgesiana, E. cypellocarpa, E. deglupta, E. fastigata, E. globoidea, E. obliqua, E. sideroxylon, E. sieberi, Eucalyptus spp.* (7), *E. tereticornis, E. vitrea, Eugenia acapulcensis, E. baileyi, E. banksii, E. botequimensis, E. brachipoda, E. brasiliensis, E. capensis, E. caryophyllata, E. confusa, E. capuli, E. corynantha, E. coryno-carpa, E. costata, E. cumini, E. effusa, E. grandis, E. inundata, E. jambolana, E. jambos, E. malaccensis, E. microcarpa f. robusta, E. microphylla, E. mirandae, E. monticola, E. natalitia, E. neurocalyx, E. patrisii, E. pluriflora, E. posoneura, E. punicaefolia, E. pyriformis, E. rapana, E. rubicunda, E. sandwicensis, Eugenia spp.* (11), *E. tierneyana, E. uniflora, E. verrucosa, E. zuluensis, Feijoa sellowiana, Fenzlia* (= *Myrtella*) *obtusa, Gomidesia* (= *Myrcia*) *anacardiaefolia, G. flagellaris, G. spectabilis, Heteropyxis natalensis, Jambosa* (= *Syzy-*

gium) kurzii, J. ramosissima, Kunzea pomifera, Leptospermum eri-coides, L. scoparium, Lysicarpus angustifolius, Mearnsia (= *Metrosideros*) *cordata, Marlieria sp., Melaleuca dealbata, M. ericifolia, M. gibbosa, M. gnidioides, M. leucadendron, M. neglecta, M. nodosa, M. quinquenervia, M. squamea, Metrosideros angustifolius, M. collina, M. macropus, M. perforata, M. robustus, M. excelsa, M. nervulosa, Metrosideros spp.* (5), *Mitranthes maria-aemiliae, Moorea* (= *Cloezia*) *sp., Myrceugenia acrophylla, M. enosma, M. myrcioides, M. regnelliana, Myrcia acuminatissima, M. bracteata, M. castrensis, M. cordifolia, M. deflexa, M. glabra, M. hatschbachii, M. heringii, M. huanocensis, M. obtecta, M. paraensis, M. paivae, M. pubipetala, M. rostrata, Myrcia spp.* (3), *M. splendens, M. sphaerocarpa, M. sylvatica, M. tomentosa, Myrciaria ciliolata, M. cuspidata, M. delicatula, M. hatschbachii, M. tenella, Myrrhinium rubiflorum, Myrtus bullata, M. communis, M. obcordata, Myrtus sp., Octamyrtus behramannii, O. pleiopetala, Pimenta dioica, P. officinalis, Pimenta sp., Plinia pinnata, Pisdium arboreum, P. cattleyanum, P. ehrenbergii, P. guajava, P. hatschbachii, P. littorale, P. sartorianum, Psidium spp.* (2), *Rhodamnia trinervia, Rhodomyrtus calophlebia, R. macrocarpa, R. novoguineensis, R. pinnatinervis, R. tomentosa, Siphoneugenia widgreaniana, Syncarpia laurifolia, Syzygium adelphicum, S. claviflorum, S. cordatum, S. cuminii, S. gerrardii, S. guineense, S. huillense, S. jambolanum, S. jambos, S. rubiginousum, Syzygium spp.* (4), *S. tetragonium, Thryptomene mainsonneurii, Tristania conferta, T. exiliflora, T. glauca, T. longivalis, Tristania spp.* (2), *Xanthostemon chrysanthus, Xanthostemon sp.*

N

NAJADACEAE
1 genus; 35 species

This is a family of cosmopolitan fresh water aquatics. Alkaloids are not known.

In the tests recorded here, neither *Najas marina* nor two unidentified species of *Najas* were positive.

NELUMBONACEAE
1 genus; 2 species

These water lilies are ornamental aquatics of eastern Asia and North America, often included earlier in the Nymphaeaceae.

The two species are alkaloidal; in the present study, six samples of *Nelumbo lutea* and five of *N. nucifera* were alkaloid-positive as has been recorded in earlier literature.

NEPENTHACEAE
1 genus; 70 species

These insectivorous pitcher plants are found from the Seychelles and Madagascar to Australia and New Caledonia. They are often cultivated as novelties.

Alkaloids have not been discovered in the family; *Nepenthes mirabilis* and two undetermined species of the genus were negative in the tests conducted here.

NYCTAGINACEAE
34 genera; 350 species

The family is found in the tropics and subtropics of both hemispheres but mainly in the New World. Its members have some use as ornamentals.

Cyanogenesis and alkaloidlike substances (betanidins) occur in the family; positive alkaloid tests have been recorded for a few genera, including *Mirabilis jalapa,* which was positive (1/5) in this study. The following were also positive: *Ceodes umbellifera, Commicarpus pentandrus, Mirabilis bigloveii* (1/2), *Neea oppositifolia.*

Negative species included the following: *Abronia latifolia, A. maritima, A. pogonatha, A. turbinata, A. umbellata, A. villosa, Acleisanthes longiflora, A. incarnata, A. pseudaggregata, Allionia sp., Boerhavia grandis, B. intermedia, B. anisophylla, B. caribaea, B. diffusa, B. erecta, B. verticillata, B. viscosa, Bougainvillea spectabilis, Ciriba sp., Commicarpus (= Boerhavia) africanus, C. fallacissimus, C. plumbagineus, C. scandens, Cyphomeris (= Boerhavia) gypsophiloides, Mirabilis laevis, M. longiflora, M. nyctagina, Mirabilis sp., M. tenuloba, Neea glomeruliflora, N. ovalifolia, N. psychotrioides, Neea spp.* (2), *Okenia hypogaea, Oxybaphus* (= *Mirabilis*) *comatus, O. glabrifolius, Phaeoptilum spinosum, Pisonia aculeata, P. brunonianum, P. cuspidata, P. diandra, P. eggersiana, P. longirostris, P. mulleriana, P. olfersiana, P. salicifolia, P. umbellifera, Pisoniella arborescens, Salpianthus arenarius, S. purpurascens, Torrubia* (= *Pisonia*) *obtusata.*

NYMPHAEACEAE
6 genera; 60 species

These are aquatics in fresh water, used as ornamentals and found worldwide. Alkaloids are likewise known and those of *Nuphar* have been described.

As expected, *Nuphar advena* (2/2), *N. luteum* (5/5), and *N. odorata* (1/2) were positive, as were *N. lotus* (1/3) and *N. stellata. Nymphaea lutea, N. capensis, N. maculata,* and an undetermined species of *Nymphaea* were negative.

NYSSACEAE
3 genera; 8 species

This is a family of eastern North America and Asia, closely related to the Cornaceae and included in that family by some botanists. Some members are used as timber, others as ornamentals.

Only the antitumor alkaloids of *Camptotheca* have been reported. *Davidia involucrata,* as tested here, was positive (1/3) (it is sometimes placed in a family of its own, Davidiaceae); three species of *Nyssa* were negative: *N. biflora, N. ogeche, N. sylvatica.*

O

OCHNACEAE
37 genera; 460 species

This family is found especially in Brazil; it yields some timber and cultivated ornamentals.

Alkaloids are not known. Of 32 species tested, positive results were obtained here for: *Luxemburgia octandra, Ochna pretoriensis, Ouratea flexousa, Schuurmania henningsii.*

Negative tests were obtained with samples of *Blastemanthus sp.*, *Brackenridgea australiana, Elvasia elvaseoides, Luxemburgia spp.* (2), *Ochna atropurpurea, O. holstii, O. leptoclada, O. longipes, O. mossambicensis, O. natalitia, O. o'connorii, O. pulchra, O. schweinfurthiana, Ochna sp., Ouratea angustifolia, O. aromatica, O. discophora, O. parviflora, O. sellowii, Ouratea spp.* (5), *Planchonella anteridifera, P. chartacea, P. costata, P. myrsinoides, P. pohlmaniana, P. toricellensis, Sauvagesia erecta, S. linearifolia, Wallacea insignis.*

OLACACEAE
29 genera; 200 species

The Olacaceae furnish part of the local spices and some medicines in tropical and southern Africa. The family is very variable and, at present, is an accumulation of plants that were previously considered included in several families.

Alkaloids or alkaloidlike compounds are known from about half a dozen genera of the family. In the present study, the following gave positive tests: *Heisteria spp.* (2), *Lirisoma macrophylla, Olax dissitiflorus, O. subscorpionoidea.*

Negative tests were obtained with the following: *Anacolosa papuana, Curupira tefeensis, Heisteria scandens, H. sessilis, Minquartia guianensis, Olax wightiana, Ptychopetalum olacoides, Tetrostylidium, Ximenia americana, X. caffra, X. parviflora.*

OLEACEAE
24 genera; 900 species

The family of the olive, known and widely used since ancient times, supplies also timber (e.g., ash) and cultivated ornamentals and shrubs (e.g., lilac, jasmine), some of which are also used in perfumery.

A few genera have given positive alkaloid tests in earlier surveys due, probably, to pseudoalkaloids of iridioid structure and to quinine-type alkaloids in *Jasminum* along with other pyridinoids.

The following known alkaloidal species gave positive tests in this survey as well: *Jasminum domatiigerum, J. schumannii, Lingustrum ovalifolium, L. sinense* (1/3), *Linociera* (= *Chionanthus*) *axillaris, Olea europaea.* These were also positive: *Chionanthus virginicus* (2/3), *Fraxinus dipetala* (1/2), *Jasminum angulare, J. didymium, J. fluminense* (1/3), *J. multipartitum, J. pseudoanastomosans, J. quinatum, Jasminum spp.* (2/4), *J. streptopus, Lingustrum japonicum* (2/5), *L. microcarpum, L. novoguinense, Linociera brassii, Linociera spp.* (4/4), *Olea capensis* (1/2), *O. exaaperata, Schrebera argyrotrichia.*

There were also a number of alkaloid-negative species: *Chionanthus retusus, C. serrulatus, Fontanesia fortunei, Fontinalis sp., Forestiera angustifolia, F. phillyreoides, F. puberula, F. racemosa, F. tomentosa, Fraxinus americana, F. berlandieriana, F. excelsior, F. greggii, F. griffithii, F. insularis, F. nigra, F. oregona, F. ornus, F. pauciflora, P. pennsylvanica, F. potosina, F. rufescens, F. uhlei, Gymnelaea lanceolata, Jasminum azoricum, J. breviflorum, J. gracile, J. humile revolutum, J. mesnyi, J. officinalis, J. primulinum, J. pubescens, J. rottlerianum, J. roxburghianum, J. simplicifolium, J. stenolobium, J. subtriplinerve, Lingustrum lucidum, L. obtusifolium, L. vulgare, Linociera foveolata, L. mandiocana, Menodora helianthemoides, M. scoparia, Notalaea microcarpa, Olea africana, O. apetala, Osmanthus americana, O. fragrans, O. ilicifolius, O. sandwicensis, O. sieboldii, Osmanthus sp., Osmaria burkwoodi, Schrebera alata, Syringa persica.*

OLINIACEAE
1 genus; 8 species

This small family of eastern and southern Africa has no economic importance.

Cyanogenesis is known but alkaloids have not been found. Three samples representing three species of *Olinia* were tested without positive results: *Olinia cymosa, O. emarginata, O. radiata.*

ONAGRACEAE
24 genera; 650 species

A cosmopolitan family, it is found principally in warm and temperate America. A familiar member, cultivated as an ornamental, is *Fuchsia*.

A few positive alkaloid tests had been recorded earlier, but little is known of the chemical nature of the compounds.

Positive results were obtained on testing *Diplandra* (= *Lopezia*) *lopezioides, Fuchsia thymifolia* (1/2), *Jussiaea decurrens* (= *Ludwigia*) (1/2), *Lopezia sp., Ludwigia palustris* (1/2), *Semeiandra* (= *Lopezia*) *grandiflora*.

The following species were negative: *Boisduvalia densiflora, Circaea quadrisulcata, Clarkia rhomboidea, C. speciosa, C. unguiculata, Epilobium billardierianum, E. bonplandianum, E. coloratum, E. hirsutum, E. hooglandii, E, keysseri, E. watsonii, Fuchsia arborescens, F. boliviana, F. coccinea, F. cylindracea, F. excortica, F. intermedia, F. michoacanensis, F. microphylla, F. minutiflora, F. parviflora, F. regia, Fuchsia spp.* (5), *F. tetradactyla, Gaura angustifolia, G. biennis, G. coccinea, G. tripelata, Gayophytum diffusum, Hauya elegans, H. haydeana, Jussiaea affinis, J. erecta, J. foliobracteolata, J. leptocarpa, J. longifolia, J. peruviana, Jussiaea sp.* aff. *peruviana, J. repens, J. sericea, Jussiaea spp.* (4), *J. suffruticosa, Lopezia hirsuta, L. pubescens, L. racemosa, L. trichota, Ludwigia alternifolia, L. ascendens, L. erecta, L. hyssopifolia, L. latifolia, L. micrantha, L. natans, L. octovalvis, L. parviflora, L. pilosa, L. prostrata, L. pubescens, L. stenorraphe, Oenothera alyssoides, O. biennis, O. caespitosa, O. cheiranthifolia, O. clauiformis, O. decorticans, O. deltoides, O. dissecta, O. drummondii, O. greggii, O. hookeri, O. humifusa, O. kunthiana, O. lacinata, O. rosea, Oenothera sp., O. speciosa, O. tanacetifolia, O. tetraptera, O. xylocarpa, Riesenbachia* (= *Lopezia*) *racemosa, Zauschneria* (= *Epilobium*) *californica, Z. latifolia*.

OPILIACEAE
9 genera; 28 species

This is a tropical family yielding a few edible fruits.

Alkaloid reactions have been obtained on testing species of *Op-*

ilia and *Rhopalopilia*, but specific compounds have been neither isolated nor identified. In this study four species of *Agonandra* were tested with negative result: *A. brasiliensis, A. obtusifolia, A. racemosa, Agonandra sp.*

ORCHIDACEAE
Over 700 genera; about 30,000 species

The orchids currently are considered to be the largest family of flowering plants. Both Willis and Mabberley cite about 17,000 species; other authorities put the number in the neighborhood of 30,000. Except for their horticultural value, only one genus, *Vanilla*, yields an important product.

Considering the size of the family, its chemistry is not that well known. Alkaloids were detected in some genera as early as 1892, and records now exist for their presence in some 800 species of 180 genera. To this record, several of the following species may now be added. With few exceptions, they were tested as herbarium specimens in the Oakes Ames Herbarium of Harvard University; in deference to the size of the family, samples were taken from every tenth sheet regardless of genus or species.

Eight of the species tested were known to contain alkaloids from previous studies: *Bromheadia finlaysoniana* (1/3), *Chysis bractescens, Compylocentrum micranthum, Lockhartia oerstedii, Maxillaria fulgens, Oncidium pumilum* (2/2), *Rodriguezia decora, Trichopilia fragrans.*

A few others were likewise positive: *Agrostophyllum obscurum, Anota* (= *Rhynchostylis*) *hainanensis, Arachnis longicaulis, Bollea coelestis, Brassia caudata, Bulbophyllum blumei, B. cochleatum, Calanthe alpina* var. *fimbricata, C. ensifolia, C. lamellosa, C. pulchra* (1/2), *Catasetum integerrinum* (2/3), *C. planiceps, C. russellianum, C. suave, Chameanthus* (= *Geogenanthus*) *wenzelii, Chondrorrhyncha lendyana, Cleisostoma* (= *Sarcanthus*) *spathulatum, Dendrobium anosum, Dendrophylax funalis, Epidendrum gladiatum* (2/2), *Eulophia angolensis, E. clitellifera, Galeandra baueri, G. beyrichii, G. devoniana* (2/2), *Liparis neglecta, Liparis sp., Lycaste virginalis, Maxillaria bracteorum, M. densa* (2/2), *M. friedrichsthalii* (1/2), *M. biolleyi, Odontoglossum pardium, Oncidium ovatilabium,*

O. paraneme, O. superbiens, O. tigrinum, Ornithocephalus tripterus, Pachyphyllum distichum, Pachyphyllum sp., Pescatoria lehmanni, Pholidota chinensis, Phreatia congesta, Plocoglottis bicallosa, Podochilus tenuis (1/3), *Polycycnis muscifera, Rangaeris rhipsalisocia, Renanthera elongata, Saccolabium luzonense, S. saxicolum, Spathoglottis chrysantha, Trichopillia sp., T. tortilia, Trigonidium egertonianum* (1/3), *Zygopetalon triste.*

Substances yielding indigo through hydrolysis of indican in the presence of oxygen during drying of the plant parts are found in several species of the family.

Of the total of 1,245 species of orchids examined, most were found to be alkaloid-negative. However, it is fair to note that many of these had been listed by others as positive, suggesting a set of false negative results due to any one factor or a combination of the factors discussed in the Preface with respect to herbarium specimens and their preservation.

Negative tests included the following: *Acampe multiflora, Acanthohippum bicolor, A. papuanum, A. martinianum, Acineta alticola, A. superba, Acriopsis javanica, A. philippensis, Acrolophia cochlearis, A. lamellata, A. tristis, Adenocos* (= *Sarcochilus*) *virens, Aerangis* cf. *somalensis, A. laurentii, Aerides jackianum, A. lawrenciae, A. lineare, A. multiflora, A. odoratum, A. quinquevulnera, A. radicosum, Aganisia cyanea, A. fimbricata, A. pulchella, Agrostophyllum bicuspidatum, A. appendiculoides, A. brachiatum, A. callosum, A. carinoides, A. confusum, A. costatum, A. denbergeri, A. graminifolium, A. hasseltii, A. inocephalum, A. javanicum, A. laxum, A. leucocephalum, A. longifolium, A. longifolium* var. *obtusifolium, A. longivaginatum, A. luzonense, A. malindangense, A. mearnsii, A. megalurus, A. mucronatum, A. palauense, A. paniculatum, A. parviflorum, A. pelorioides, A. philippinensis, A. saccatum, A. spicatum, A. stipulatum, A. sumatranum, A. superpositum, A. uniflorium, A. wenzelii, A. zeylanicum, Ancistrochilus rothschildiana, Angraecum imbricatum, A. birrimense, A. conchiferum, A. distichum, A. giryanae, A. infudibulare, A. schollerianum, A. subulatum, Anguloa virginalis, Ansellia gigantea, A. gigantea* var. *nilotica, Anthogonium gracile, Aplectrum hymenale, Appendicula alba, A. anceps, A. anemophila, A. angustifolia, A. bifaria, A. bracteosa, A. buxifolia, A. callosa, A. cleistogma, A. clemensiae,*

*A. congenera, A. cornuta, A. crotalina, A. cuneata, A. dendro-
boides, A. effusa, A. elmeri, A. fenixii, A. foliosa, A. grandiflora,
A. infundibuliformis, A. irigensis, A. kinabaluensis, A. bracteata,
A. latibracteata, A. longirostrata, A. lucbanensis, A. lucida, A. luzo-
nensis, A. malindangensis, A. maquilingensis, A. magnibracteata,
A. micrantha, A, muricata, A. ovalis, A. pauciflora, A. pendula,
A. philippinensis, A. polyantha, A. pseudopendula, A. ramosa,
A. reflexa, A. torta, A. unciferus, A. undulata, A. vanikorensis,
A. weberi, A. xytriophora, Arachnanthe (= Arachnis) julingii,
Arachnis annamensis, A. clarksi, A. longicaulis, Ascocentrum au-
riantiacum, Ascotainia (= Tainia) elmeri, Aspasia epidendroides,
A. principessa, A. variegata, Bifrenaria aurantiaca, B. aurea,
B. harrisoniae, B. longicornis, B. sabulosa, Bletia catenulata,
B. florida, B. gracile, B. palmeri, B. patula, B. purpurea, B. reflexa,
B. tuberosa, B. wagneri, Brassia allenii, B. bidens, B. chlorops,
B. gireondiana, B. longissima, B. maculata, B. verrucosa, B. wag-
neri, B. warszewiczii, Bromheadia alticola, B. borneensis, Bulbo-
phyllum adenopetalum, B. affine, B. africanum, B. amatum, B. ama-
nicum, B. angustifolium, B. antenniferum, B. austatum, B. baileyi,
B. banthanthum, B. bataanense, B. bolaninum, B. bufo, B. calama-
ria, B. capitatum, B. coelogyne, B. coneinum, B. crassicaudatum,
B. croceum, B. cumingii, B. cupreum, B. dasypetalum, B. dearii,
B. delitescens, B. distans, B. drymoglossum, B. dulitense, B. ebracteatum, B. eladium, B. endotrachys, B. eximum, B. falcatum,
B. foseum, B. gracile, B. grandiflorum, B. guamense, B. harposepa-
lum, B. hastatum, B. hymenobracteatum, B. igneum, B. imbricatum,
B. inconspicum, B. jaequetii, B. kwangtungense, B. lanceolatum,
B. linderi, B. longiflorum, B. magnivaginatum, B. makaya-
num, B. maximum, B. medusea, B. membranifolium, B. minday-
aense, B. montense, B. nageli, B. nebulosum, B. nuruanum, B. nyas-
sum, B. ochroleucum, B. ovalifolium, B. oveogenum, B. oveonastes,
B. oxypetalum, B. pachyneuron, B. pachyrrhacis, B. pallidiflorum,
B. parvilobium, B. patens, B. pergracile, B. phacopogon, B. pobe-
guini, B. ponapense, B. popayanense, B. praealtum, B. preticei,
B. profusum, B. puberulum, B. purpurascens, B. pygmaeum,
B. radiatum, B. rhizomatosum, B. solaceense, B. schizopetalum,
B. sigmoidenum, B. smithanthum, B. tahitense, B. tibeticum,
B. trimeni, B. trigetum, B. uniflorum, B. vaginatum, B. velutinum,*

B. volkensii, B. vulcanicum, Calanthe alta, C. angustifolia, C. arcuata, C. arisanensis, C. buccinifera, C. ceciliae, C. clavata, C. conspicua, C. corymbosa, C. crumenata, C. curculioides, C. davidii, C. delavayi, C. densiflora, C. discolor, C. emarginata, C. engleriana, C. ensifolia, C. flava, C. foestermannii, C. furcata, C. gracillima, C. gracilis, C. halconensis, C. hattorii, C. hennsii, C. henryi, C. hololenca, C. japonica, C. lacerata, C. lineariloba, C. liukiuensis, C. lyroglossa, C. maquilingensis, C. masuca, C. macgregorii, C. megalopha, C. mexicana, C. nephroglossa, C. plantaginea, C. puberula, C. purpurea, C. reflexa, C. rhodochila, C. schliebenii, C. speciosa, C. striata, C. sylvatica, C. tahitensis, C. tricarinata, C. vaupelliana, C. ventilabrum, C. veratrifolia, C. volkensii, Calyptrochilum preussii, C. christyanum, Camaridium (= Maxillaria) ochroleneum, Camarotis papuana, C. philippinensis, C. brenesii, C. jamaicense, C. carrettiae, Campylocentrum neglectum, C. outonei, C. tuerckheimii, C. schiedei, Catasetum confiformes, C. dilectum, C. discolor, C. interrimum, C. laminatum, C. maculatum, C. macrocarpum, C. oerstedii, C. saccatum, C. tabulare, C. viridoflorum, C. warscewiczii, Cattleya aurantiaca, Cephalangium sp., Cephalangraecum (= Ancistrorhynchus) capitatum, Ceratostylis caespitosa, C. capitata, C. dischorensis, C. flavescens, C. gracilis, C. grandiflora, C. kaniensis, C. latifolia, C. latipetala, C. leucantha, C. loheri, C. malaccensis, C. micrantha, C. philippinensis, C. ramosa, C. rubra, C. scirpoides, C. senilis, C. simplex, C. subulata, C. teres, C. wenzelii, Chamaeangis odoratissima, Cheiradenia imthurnii, Chilopogon (= Appendicula) bracyeatum, C. distichum, C. kinabaluensis, Chitonanthera aporoides, C. brassii, C. gracilis, C. lorentzii, C. oberonoides, C. tenuis, Chondrorhyncha chestertoni, C. albicana, C. endresii, C. lipscombia, Chrysocycnis rhomboglossum, C. schlimii, Chysis aurea, Cleisostoma (= Sarcanthus) expansum, C. maculosum, Comparettia falcata, Corallorhiza macrantha, C. maculata, C. mertensiana, C. odontorhiza, C. striata, C. trifida, C. wisteriana, Coryanthes elegantium, C. macrantha, C. speciosa, Cremastra triloba, Cryptarrhena lunata, Cryptocentrum calearatum, Cryptochilus sanguinea, Cyanaerochis arundinae, Cymbidium alifolium, C. angustifolium, C. atropurpureum, C. devonianum, C. elegans, C. ensifolium, C. faberi, C. finlaysonianum, C. floribundum, C. grandiflorum, C. longifo-

lium, C. mackinnoni, C. pendulum, C. pseudovirens, C. tracyanum, C. ustulatum, C. virescens, C. yunnanensis, Cynorkis egertonianum, C. stelliferum, C. ventricosum, C. warscewiczii, Cyprepedium acaule, C. irapeanum, Cyritidum (= Cyrtorchis) buchtienii, Cyrtopodium broadwayi, C. cristatum, C. paraense, C. punctatum, C. purpureum, Cyrtorchis montierae, C. praetermissa, Dendrobium sp., Dendrobium acuminatissimum, Diaphananthe fragrantissima, D. bidens, Diadenium micranthum, Dichaea ciliolata, D. dammeriana, Dichaea echinocarpa, D. glauca, D. graminoides, D. histricina, D. lankesterii, D. morrisii, D. muricata, D. panamensis, D. pendula, D. picta, D. powellii, D. trulla, D. willdenowiana, Diploprora championii, Dipodium ensifolium, D. paludosum, Doritis philippinensis, Earina bronsmickii, E. deplanchei, E. laxior, E. mucronata, E. plana, E. valida, Elleanthus capitatus, Epiblastus acuminatus, E. merrillii, E. schultzei, E. ornithioides, Epidendrum cochleatum, E. difforme, E. gonioehachis, E. ibaguense, E. mosenii, E. musicicola, E. ochraceum, E. paranthicus, E. radicans, Epidendrum spp. (5), E. varicosum, Epipactus gigantea, Eria senilis, E. taylori, E. vanoverberghii, E. ventricosa, E. vulpina, E. whitfordii, E. woodiana, E. zamboagensis, Eriopsis biloba, E. sprucei, Erycina echinata, Eulophia sp., E. squalida, E. stachyodes, E. stenophylla, E. stricta, E. virens, E. zeyheri, Eulophidium (= Oeceoclades) alta, E. angolensis, E. beravensis, E. caffra, E. calanthoides, E. caricifolia, E. clavicorius, E. cochlearis, E. compestris, E. cucullata, E. dahlianum, E. dregiana, E. ecristata, E. ensata, E. gracilis, E. guineensis, E. hildebrandtii, E. hormusjii, E. horsfallii, E. livingstoniana, E. longifolia, E. lotilabius, E. macgregorii, E. mackinnoni, E. macrostachya, E. maculatum, E. micrantha, E. nuda, E. ovalifolia, E. paiveana, E. petersii, E. poilanei, E. pulchrum, E. schimperiana, Eulophia sp., Finetia (= Neofinetia = Holcoglossum) flacota, Galeandra baueri, G. beyrichii, G. devoniana, G. graminoides, G. juncea, G. junceoides, G. paranaensis, G. pubicentrum, Gastrochilus calceolaris, Geodorum citrinum, G. dilantatum, G. nutans, G. pictum, G. purpureum, Glomera bambusiformis, G. erythrosma, G. keysseri, G. macdonaldii, G. montana, G. rugulosa, G. schultzei, G. stenocentron, G. tenuis, Gongora cassidea, G. galeata, G. maculata, G. truncata, G. unicolor, Goodyera oblongifolia, G. procera, Govenia liliacea, G. nutica,

G. superba, G. utriculata, Grammatophyllum speciosum, G. scriptum, G. multiflorum, Gussonea (= Solenangis) chilochistae, Habenaria clypeata, H. entomantha, H. guillemini, H. repens, Habenaria spp. (6), *H. strictissima, Hexalectris brevicaulis, H. grandiflora, Houlletia brockelhurstiana, H. roraimensis, Huntleya burtii, H. lucida, H. meleagris, Hybochilus inconspicuus, Ione (= Sunipia) andersoni, Ionopsis paniculata, I. satyroides, I. utricularioides, Kefersteinia lactea, K. parvilabris, K. pulchella, K. tolimensis, Kingiella (= Kingidium) decumbens, Koellensteinia graminea, Lacaena spectabilis, Laelia speciosa, L. carinatus, Lemurorchis madagascariensis, Leochilius gracilis, L. oncidioides, L. pygmaeus, L. tricuspidatus, Leochilus tuerckheimii, Lissochilius (= Eulophia) antennisepalus, L. arenarius, L. krebsii, L. olivcrianus, L. porphyroglossus, L. ruwenzoriensis, L. sceptum, L. validus, L. wakefieldii, Listrostachys bidens, L. pellucida, Lockhartia hercodonta, L. micrantha, L. pallida, L. amoena, L. elegans, Luisia foxworthyi, L. ramosii, L. tenuifolia, L. teretifolia, L. trichorhiza, Lycaste aromatica, L. barringtoniae, L. campbellii, L. candida, L. costata, L. cruenta, L. deppei, L. dowiana, L. gigantea, L. locusta, L. macrophylla, L. powellii, L. xytriophora, Macradenia brasavolae, M. lutescens, Malaxis chreubergii, Malleola constricta, M. palustris, Maxillaria acicularis, M. acuminata, M. affinis, M. aggregata, M. alba, M. alticola, M. ampliflora, M. anceps, M. angustissima, M. appendiculoides, M. arachnitiflora, M. aurea, M. brachypetala, M. brachybulbon, M. bracteatum, M. brevilabia, M. camaridii, M. campanulata, M. connellii, M. colorata, M. coccinea, M. conferta, M. cobanensis, M. crassifolia, M. crassicaulis, M. ctenostachys, M. curtipes, M. devauxiana, M. densiflora, M. dendroboides, M. diviniflora, M. disticha, M. diburna, M. elatior, M. elongata, M. endresii, M. eueullata, M. exaltata, M. fasciculata, M. falcata, M. ferdinandina, M. floribunda, M. flava, M. fractiflexa, M. funera, M. fucata, M. graminifolis, M. histrionica, M. inaudita, M. lepidota, M. linearifolia, M. longibracteata, M. mapiriensis, M. marjuata, M. maleolens, M. meridensis, M. maleagris, M. microphyton, M. minus, M. nanegalensis, M. nagelii, M. nasuta, M. nioea, M. notylioglossa, M. oreocharis, M. paribulbosa, M. pastensis, M. pauciflora, M. parvilabia, M. palleatum, M. plemanthoides, M. planicola, M. powellii, M. pulchra, M. punctostriata, M. purpu-*

rea, M. ramonensis, M. reichenheimina, M. ringens, M. rufescens, M. sanguinea, M. sanguinolenta, M. scurrilis, M. serrulata, M. sigmoidea, M. sophronitis, M. spilotantha, M. stenophylla, M. striata, M. superflue, M. tafallae, M. tenuifolia, M. tonduzii, M. trigona, M. umbratilis, M. uncata, M. urbaniana, M. valenzuelana, M. variabilis, M. vandiformis, M. violacea, M. villosa, M. wercklei, M. wrightii, M. xantholenea, M. xylobiiflora, Mediocalar crenatulum, Mediocalar sp. cf. *doctersii, M. monticola, M. pygmaeum, M. paradoxum, M. robustum, M. siphyllum, M. uniflorum, M. vanikorense, Menadenium (= Zygosepalum) labrosum, Mesospinidium warscewiczii, Microdelia exilis, Microsaccus longicalcaratus, Microtis uniflora, Miltonia endersii, M. flavescens, Mormodes aromatica, M. atropurpurum, M. maculata, Mormolyca polyphylla, M. rigens, Mystacidum distichum, M. xanthopollinium, Nageliella purpurea, Neodenthamia gracilis, Neodryas rhodoneura, Notylia barkeri, N. bicolor, N. buchtienii, N. latilabia, N. replicota, N. sagittifera, N. sylvestris, N. venezuelana, Octarrhena amesiana, O. angraecoides, O. caulescens, O. gemmifera, O. parvula, Odontoglossum angustifolium, O. bictoniense, O. brachypterum, O. brevifolium, O. cervantesii, O. convallarioides, O. cordatum, O. egertonii, O. grande, O. hallii, O. laevis, O. maculatum, O. mirandum, O. nebulosum, O. oerstedii, O. pendulum, O. polyxanthum, O. pulchellum, O. ramosissimum, O. reichenheimii, O. rossii, O. schlieperianum, O. stellatum, O. stenoglossum, Oncidium ampliatum, O. ansiferum, O. armillare, O. aureum, O. ascendens, O. baueri, O. bicallosum, O. bifolium, O. blanchetii, O. brachyandrum, O. bryolophotum, O. cabagrae, O. cavendishianum, O. ceballota, O. chartaginense, O. cheirophorum, O. chrysopterum, O. crispum, O. cristagalli, O. falcipetalum, O. flexuosum, O. globuliferum, O. hartatum, O. heteranthum, O. hyphoemalicum, O. incervum, O. intermedium, O. isthmii, O. johannis, O. leucochilum, O. liebmannii, O. longifolium, O. luridum, O. macranthum, O. maculatum, O. montanum, O. nigratum, O. nubigenum, O. oblongatum, O. obryzatum, O. ochmatochilum, O. ornithorhynchum, O. panamense, O. panduriforme, O. pergameneum, O. pittieri, O. polycladium, O. polyodenium, O. powellii, O. pulchellum, O. pusillum, O. ramiferum, O. reflexum, O. refractum, O. retusum, O. sarcodes, O. sphacelatum, O. stenotis, O. stipitatum, O. tetrapetalum, O. titania,*

O. triquetrum, O. trulla, O. tuerckheimii, O. uniflorum, O. urophyllum, O. variegatum, O. ventilabrum, O. viperinum, O. warscewiczii, O. wentworthianum, O. wydleri, O. zebrinum, Ornithidium (= Maxillaria) chloroleucum, *O. coccinium, O. densum, O. proliferum, Ornithidium sp.* cf. *anceps, O. vestitum, Ornithocephalus bicornis, O. gladiatus, Ornithochilus fuscus, Otostylis brachystalix, Oxyanthera (= Thelasis) elata, Pachyphyllum hartwegii, P. squarrosum, P. muscoides, Peristeria elata, P. pendula, Pescatoria cerina, Phaius tankervillii, Phalaenolepis petoletii, P. ludemanniana, P. equestris, P. amabilis, Phreatia amesii, P. aristulifera, P. cauligera, P., collina, P. densiflora, P. luzonensis, P. matthewsii, P. microphyton, P. myosuriforme, P. obtusa, P. petiolata, P. pusilla, P. reineckei, P. samoensis, P. secunda, P. semiorbicularis, P. sphaerocarpa, P. stenostachya, P. sulcata, P. thomsonii, P. upolensis, P. urostachya, P. vanoverberghii, P. yunkeri, Pilophyllum laricinum, Plantanthera maridosimorum, Pleurothallis grobyi, P. arenata, Plocoglottis acuminata, P. copelandii, P. foetida, P. javanica, P. lucbanensis, P. maculata, P. mindorensis, Poaephyllum parviflorum, Podochilus bimaculatus, P. cultratus, P. cumingii, P. folcatus, P. intricatus, P. longilobus, P. luscens, P. malabaricus, P. microchilus, P. plumosus, P. saxatilis, P. scalpelliformis, P. sciuroides, P. serpyllifolius, P. similis, P. uncata, P. viellardii, Polycycnis lehmanii, P. vittata, Polyrrhiza (= Dendrophylax) gracilis, P. linderii, Polystachya affinis, P. cerea, P. clavata, P. colombiana, P. cultriformis, P. dendrobiiflora, P. dolichophylla, P. estrellensis, P. foliosa, P. fusiformis, P. holstii, P. imbricata, P. keimsiana, P. laxiflora, P. leonensis, P. lineata, P. luteola, P. masayensis, P. minuta, P. nigrescens, P. odorata, P. ottoniana, P. retusiloba, P. rhodoptera, P. shega, P. simplex, P. spetalla, P. stuhlmannii, P. tayloriana, P. villosa, P. vulcanica, P. zambesiaea, Pomatocalpa bicolor, P. densiflora, P. latifolium, P. vitellinum, Pterygodium caffrum, Quekettia micromera, Renanthera alba, R. bilinguis, R. coccinea, R. imschootiana, R. matutina, R. philippinensis, R. storiei, Rhipidoglossum (= Diaphananthe) rutilum, Rhynchostylis retusa, R. violacea, Robiquetia pantherina, R. leuta, R. merrillii, R. vanoverberghii, Rodriguezia candida, R. batemannii, R. decora, R. epiphyta, R. lehmannii, R. microphylla, R. secunda, R. refracta, Saccolabium calceolare, S. constrictum, S. chrisiflorum, S. distichum, S. fili-*

forme, S. gramense, S. miniatum, S. miserum, S. roseum, S. succisum, Sarcanthus (= Cleisostoma) arevipes, S. bicuspidatus, S. bifidus, S. clemeniae, S. dealbatus, S. elongatus, S. insectifer, S. merrillianus, S. micranthus, S. pachyphyllus, S. rostralis, S. striolatus, S. subulatus, S. turbineus, S. utriculosus, S. weberi, Sarcochilus appendiculatus, S. calceolus, S. emarginatus, S. falcatus, S. hystrix, S. japonicus, S. leytensis, S. longicalcarus, S. moorei, S. pallidus, S. philippinense, Sarcochilus sp., S. teres, S. uniflorus, Sarcostoma javanica, Satyrium carneum, S. corrifolium, S. membranaceum, S. nephalense, Scaphyglottis livida, S. ocellatum, Scelochilus ottonis, Schoenorchis nicrantha, S. juncifolia, S. vanoverberghia, S. gemata, S. densiflora, Sepalosaccus humilis, Sigmatostalix brachyscion, S. hymenantha, S. guatemalensis, S. radicans, Sobralia sp., Solenangis clavata, S. scandens, Spathoglottis aurea, S. carolinensis, S. fortunei, S. micromisiaca, S. pacifica, S. plicata, S. vieillardii, Spathoglottis sp., S. tomentosa, Spiranthes aurantiaca, S. cinnabarina, S. spiralis, Stanhopea costaricensis, S. ecornuta, S. graveolens, S. grandiflora, S. haseloviana, S. hernandesii, S. intermedia, S. oculata, S. randii, S. saccata, S. wardii, Staurochilus fasciatus, Stauropsis (= Trichoglottis) fasciata, Stelis puberula, Stenorrhynchos paraguayensis, Stolzia nyssana, Symphoglossum sanguinum, Systeloglossum costaricense, S. acuminatum, Taeniophyllum fascicola, Taeniophyllum sp., Telipogon angustifolium, T. endresianum, T. gracilipes, T. phalaena, Thecostele elmeri, T. zollingerii, Thelasis carinata, T. elongata, T. micrantha, T. triptera, Theodorea (= Rodrigueziella) gomezoides, Thrixspermum agusaneme, T. amplexicaule, T. arachnites, T. centipeda, T. comans, T. graeffei, T. hainanense, T. longipilosum, T. trichoglottis, T. wenzelii, Toeneophyllum sp., Trichocentrum candidum, T. capitatum, T. hoegei, T. maculatum, T. panduratum, T. pfavii, Trichoglottis bataaensis, T. brachiata, T. fasciata, T. guibertii, T. ionosma, T. lanceolaria, T. latisepala, T. lehmannii, T. logeriana, T. luzonensis, T. magnicallosa, T. mindanaemsis, T. perezii, T. philippinensis, T. retusa, T. rizalensis, T. rosea, T. tenuis, Trichopilia fragrans, T. laxa, T. leucoxantha, T. marginata, T. rostrata, T. suavis, T. turialbae, Tridactyle anthomaniaca, T. linearifolia, T. tridactylites, Trigonidium egertonianum, T. equitones, T. latifolium, T. lankesteri, Trizeuxis falcata.

OROBANCHACEAE
17 genera; 230 species

This is a family especially of the northern hemisphere and the subtropics of the Old World. It has no real economic importance.

Reports of positive alkaloid tests in three genera have appeared, but in this study the five samples tested were negative: *Aeginetia indica, Conopholis americana, Epifagus virginiana, Orobanche fasciculata, O. ludoviciana.*

OXALIDACEAE
8 genera; 575 species

With tropical and a few temperate species, this family furnishes some edible fruits and tubers as well as ornamentals and several weeds.

Alkaloids are not known; the accumulation of oxalates is common. Twenty-eight species were tested without a positive result: *Averrhoa carambola, A. bilimbi, Biophytum abyssinicum, B. dendroides, B. sensitivum, Monoxalis robusta, Oxalis alpina, O. barrellieri, O. bowiei, O. cernua, O. corniculata, O. decaphylla, O. europhea, O. goniorrhiza, O. grayi, O. lawsonii, O. neaei, O. occidentalis, O. pes-caprae, O. pringlei, O. repens, O. sepium, Oxalis spp.* (6).

P

PALMAE
198 genera; 2,600 species

Tropical and warm areas support palms. Only a few occur in Africa.

In many areas of the world, parts of the entire plant serve many different purposes: food (coconut, sago), waxes, fruits, timber, thatch, etc. A number are familiar as cultivated ornamentals.

Positive alkaloid tests have been reported for several genera of the family, which is rich in potential biodynamic constituents. Many of the nitrogen-containing compounds have been found in the seed, which was not tested in the experiments reported here.

Hyphaene crinata and *Phoenix reclinata* gave positive tests; the remainder of the species tested were negative: *Areca catechu, Arecastrum romanzoffianum, Astrocaryum aculeatissimum, Attaleya dubiua, Bactris mexicana, Brahea dulcis, Calamus australis, C. margaritae, Caryota urens, Chamaedorea humilis, C. lindeniana, Coccothrinax argentea, Desmoncus chinantlensis, Diplothemium campestris, Elaeis guineensis, Erythea pimo, Lepidocaryum sp., Nypa fruticans, Plectocomiopsis sp., Pritchardia sp., Reinhardtia elegans, R. gracilior, Rhopalostylis sapida, Sabal bermudana, S. palmetto, Serenoa repens.*

PANDANACEAE
3 genera; 675 species

The family is found in the Old World tropics and as far south as New Zealand. The leaves have a number of local uses (e.g., thatch, cloth) and some fruits are edible.

Occasional positive tests for alkaloids have been noted earlier in the family. Here, of 22 samples representing 18 species, only three gave positive tests: *Galearia celebica, Microdesmis puberula* (branches), and *Pandanus tectorius* (1/3).

The following were negative: *Freycinetia arborea, F. banksii, F. demissa, Freycinetia spp.* (3), *F. storkii, Pandanus forsteri, P. odoratissimus, Pandanus spp.* (5), *P. thurstonii.*

PAPAVERACEAE
23 genera; 210 species

This is a well-known north temperate family if for no other reason than as the source of opium described in the earliest historical writings. It supplies many cultivated ornamentals as well as seeds that are used in some parts of the world as a source of edible oil.

The entire family is alkaloidal; known plants also found to be alkaloidal in this screening program include: *Adlumia fungosa, Argemone alba, A. grandiflora, A. mexicana* (7/8), *A. munita, Che-*

lidonium majus, Corydalis cava (root), C. incisa, Eschscholtzia californica, Hunnemannia fumariaefolia (4/4), Papaver aculeatum, Sanguinaria canadensis. In addition, the following were positive: Argemone corymbosa, A. ochrolenca, Bocconia arborea (2/3), B. frutescens (2/2), Bocconia sp., Dendromecon rigida, Dicentra scandens, Eschscholtzia capitosa, Fumaria sp. (this genus is sometimes placed in a family of its own, Fumariaceae), Sanguinaria sp.

Two samples that should have given positive tests based on literature information did not do so at the time of collection: Corydalis cava (leaf and stem) and C. glauca.

PASSIFLORACEAE
18 genera; 530 species

Tropical and warm temperate areas are the ranges of this family, especially in the Americas. Some species yield edible fruits, others are ornamentals, and a few have been used medicinally.

Alkaloids have been reported particularly in Passiflora, where harman has been identified in several species. The following gave positive tests as to be expected from earlier reports: Passiflora edulis (2/4), P. foetida (3/9), P. laurifolia.

Other positive species included Adenia digitata, Passiflora spp. (6/26), P. vespertillio, Tacsonia (= Passiflora) manicata.

Negative tests were given by: Adenia gummifera, A. senensis, Adenia sp., Dilkea johamesii, Mitostoma glaziovii, Passiflora alata, P. caerulea, P. coccinea, P. filipes, P. foetida lanuginosa, P. haematostigma, P. jileki, P. membranacea, P. mexicana, P. quadrangularis, P. rubia, P. serratifolia, P. serrato-digitata, P. setana, P. sexflora, P. speciosa, P. suberosa, P. viridiflora, Tryphostemma apetalum, T. hanningtonianum, T. sandersoni, T. viride.

PEDALIACEAE
18 genera; 95 species

This family is found in warm and tropical areas especially along coasts; some species are aquatic, and some are of importance as oilseeds (sesame) and edible fruits (*Proboscidea*). The family is placed in the Martyniaceae by some authorities.

Positive tests were obtained for *Harpagophytum peglerae* and *Pterodiscus luridus*; alkaloids are not otherwise known in the family.

Negative tests were obtained with the following species: *Ceratotheca sesamoides, C. triloba, Dicerocaryum zanguebaricum, Harpagophytum procumbens, Orgeria longifolia, Pterodiscus sp., Rogeria longiflora, Sesamothamnus guerichii, S. lugardii, Sesamum alatum, S. indicum, S. triphyllum.*

PENAEACEAE
7 genera; 21 species

The home of the family is the Cape area of southern Africa. Some species are used locally as medicinals.

Alkaloids are not known in the family. No positive tests were obtained with *Brachysiphon rupestris, Panaea acutifolia, P. mucronata, P. myrtoides, P. ovata, Sarcocolla formosa,* and *Saltera sarcocolla.*

PHILYDRACEAE
4 genera; 5 species

The family extends from Australia through southeast Asia up to southern Japan. Neither alkaloids nor economic uses are known.

Philydrum lanuginosum was negative for alkaloids in this study.

PHYTOLACCACEAE
18 genera; 65 species

The tropical and warm areas, especially those of the Americas, are home to this family. Some of the genera previously included in it have been shifted into other families, but they are included in the results cited below. Other than a few ornamentals and the use of *Phytolacca americana* leaves as a potherb, the family has little economic importance.

Alkaloid tests have been given by a few plants in this family; those may have been due to the presence of betacyanins and betaxanthins, which are not always considered as alkaloids in the strictest sense. Positive tests were given here by a few species known to contain alkaloids: *Phytolacca americana* (2/7), *P. icosandra* (1/4), *P. octandra* (2/4), and *Rivinia humilis* (1/3).

Codonocarpus cotinifolius and *Didymotheca thesioides* (both now assigned by some taxonomists to Gyrostemonaceae) as well as *Gisekia africana* (now placed in Aizoaceae) were likewise positive.

The remaining species tested were negative: *Achatocarpus nigricans, Codonocarpus attenuatus, Gallesia cororema, Gisekia pharnaceoides, Limeum sp., L. sulcatum, L. viscosum, Microtea debilis, Petiveria hexaglochin, Phaulothamnus spinescens, Phytolacca acinosa, P. decandra, P. dioica, P. dodecandra, P. f. monstruosa, P. heptandra, P. rivinioides, Phytolacca spp.* (3), *P. thyrsifolium, Seguieria americana, S. guaranticia, Seguieria sp., Semonvillea fenestrata, Stegnosperma halmifolia.*

PINACEAE
9 genera; 194 species

Many of the trees often thought of as pines actually belong in other botanical families. This family is a north temperate one extending south to central America and the West Indies with some representation in Sumatra and Java. It is well known as a source of lumber, ornamentals, paper pulp, edible seeds, and naval stores (resins and turpentine).

Occasional reports of alkaloids and amino acids have appeared with respect to species of *Abies* and *Pinus*, but the family is best known for its assortment of terpenoids.

A few earlier-known positives were encountered here: *Pinus armandii, P. coulteri, P. monophylla,* and *P. sabiana,* as well as 4/7 samples of *Tsuga canadensis.*

The rest of a total of 48 species tested were negative: *Abies concolor, A. fraseri, A. guatamalensis, A. kawakamii, A. magnifica, Cedrus atlantica, C. deodara, Larix decidua, L. laricina, Picea breweriana, P. obovata, P. rubens, Pinus aristata, P. bungeana, P. caribaea, P. cembroides, P. clausa, P. contorta, P. formosana,*

P. glabra, P. lambertiana, P. luchuensis, P. messiniana, P. mono-phylla, P. morrisonicola, P. muricata, P. murrayana, P. pinaster, P. quadrifolia, P. radiata, P. reflexa, P. resinosa, P. rigida, P. sylves-tris, P. taiwanensis, P. torreyana, P. virginiana, Pseudotsuga macro-carpa, P. menziesii, Tsuga caroliniana, T. chinensis, T. chinensis formosana, T. heterophylla.

PIPERACEAE
14 genera; 1,940 species

This is a tropical family that furnishes a wide variety of peppers used as food, spices, stimulants, folk medicines, and often house plants.

In addition to the amides responsible for the sharp taste of peppers, alkaloids of various types including aporphines have been found in the family.

The following species known to give positive alkaloid tests were likewise found to do so in this study: *Piper amalago, P. methysti-cum, P. peepuloides.* Further positive tests were obtained with *Peperomia retusa, P. umbilicata, Piper sp.* aff. *amalago, P. arboreum* (1/2), *P. betle, P. gaumeri, P. guineense, P. hamiltonii.*

Negative tests were obtained with the following: *Macropiper excelsum, Ottonia corcovadensis, O. martiana, O. ovata, Ottonia sp., Peperomia arabica, P. dindygulensis, P. hernandiifolia, P. mach-rostachya, P. pellucida, P. quadrifolia, Peperomia sp., P. urvilliana, Piper banksii, P. berlandieri, P. botogense, P. brachyrachis, P. ca-pense, P. cernuum var. grabricaule, P. cernuum, P. cordovanum, P. diancrum, P. dilatatum, P. fatokadsura, P. falculispicum, P. gau-dichaudianum, P. geniculatum, P. graeffei, P. guayranum, P. hispi-dum, P. iquitosense, P. leucanthum, P. longum, P. marginata, P. misatlense, P. nigrum, P. peltatum, P. sarmentosum, P. spectabi-lis, P. tuberculatum, P. uhedi, P. variegatum, Pothomorphe peltata, P. umbellata, Sarcorhachis obtusa.*

PITTOSPORACEAE
9 genera; 240 species

This Old World family of the warm and tropical areas, especially in Australia, has two genera in Malaysia.

The oils have been used in compounding fragrances; alkaloids have been reported but apparently have not been characterized.

Pittosporum ferrugineum (1/9) and *P. pentandrum* were alkaloidal by the tests used here, the former having been reported positive in earlier literature. The following were negative: *Billardiera cymosa, Citriobatus spinescens, Pittosporum brackenridgei, P. bracteolatum, P. crassifolium, P. daphniphylloides, P. erioloma, P. eugenioides, P. glabrum, P. oligocarpum, P. pentandrum, P. phylliraedioides, P. pullifolium, P. ramiflorum, P. sinuatum, Pittosporum spp.* (3), *P. subcatum, P. tenuifolium, P. tobira, P. umbellatum, P. undulatum, P. viridoflorum.*

PLANTAGINACEAE
3 genera; 255 species

This is a cosmopolitan family of which one genus, *Plantago*, yields the familiar laxative, psyllium seed.

Iridoid and other pseudoalkaloids have been found in several species but often in very small amounts and in the seeds or wood rather than in the leaves and stems.

Plantago lanceolata, reported to be alkaloidal, was also found so (1/2) here, along with one of three samples of *P. major*. The remainder of the species tested were negative: *Plantago asiatica, P. aundensis, P. catherinae, P. coronopus, P. depauperata, P. dregeana, P. galeottiana, P. insularis, P. lanceolata, P. major* var. *kimurae, P. media, Plantago spp.* (5), *P. tolucensis, P. varia, P. virginica.*

PLATANACEAE
1 genus; 6-7 species

This is a northern hemisphere family of "plane trees" used as lumber and as street trees in cities in spite of the untidiness they create due to their flaking bark.

Alkaloids are not known in the family nor were they found in testing *Platanus acarifolia, P. occidentalis,* and *P. racemosa.*

PLUMBAGINACEAE
22 genera; 440 species

The family is cosmopolitan and occurs especially near the sea. Positive alkaloid tests have been recorded for the wood of *Armeria* and *Plantago* and for the root of *Statice*. These are likely quaternary bases–choline and its relatives. However, no positive tests were obtained using the leaves and stems of the following: *Armeria formosa, Ceratostigma plumbaginoides, Dyerophytum africanum, Limonium brasiliense, L. carolinianum, L. dregeanum, L. kraussianum, L. linifolium, L. scabrum, L. scabrum avenaceum, M. dispersa, Plumbago auriculata, P. capensis, P. larpentiae, P. pulchella, P. zeylanica, Statice (= Armeria) sinensis.*

PODOCARPACEAE
12 genera; 155 species

The family is widely distributed, found mostly in the southern hemisphere but in Asia north to Japan, in the mountains of tropical Africa, and also in Central America. Its minor economic importance depends on its use as timber and ornamentals.

The family is not known for alkaloids; two positive tests were obtained during the course of this study: *Podocarpus falcatus* and *P. spicatus*. Twenty-three other species of the genus were negative: *P. affinis, P. amarus, P. archboldii, P. blumei, P. compactus, P. dacrydioides, P. elatus, P. elongatus, P. ferruginoides, P. gnidioides, P. hallii, P. henkellii, P. lambertii, P. latifolius, P. macrophyllus, P. miljanius, P. nagaia, P. neriifolius, P. nivalis, P. sellowii, Podocarpus sp., P. totara, P. wallichiana.*

PODOSTEMACEAE
50 genera; 275 species

A family mostly of tropical Asia and America, the Podostemaceae yield a few local foods (salad greens), and the Amazon Indians prepare a salt from the leaves of *Rhyncolacis nobilis*. Alkaloids have not been reported.

Two samples of *Torrenticola queenslandica* were negative.

POLEMONIACEAE
20 genera; 275 species

This is a family of western and northern America and Eurasia. Other than a few cultivated ornamentals, it has no economic importance.

Positive alkaloid tests have been recorded for three genera in the family, but the compounds have not yet been isolated and characterized. In this study, the following gave positive tests: *Bonplandia geminiflora* (1/2), *Eriastrum wilcoxii* (1/2), *Gilia capitata* (1/2), *G. coronopifolia*, *G. rigidula* (1/3) (previously reported), *Phlox maculata, Polemonium foliosissimum*.

Negative tests were obtained for the following: *Allophyllum divaricatum, Collomia linearis, Eriastrum densiflorum, E. plurifolium, Gilia latiflora, G. ophthalmoides, G. splendens, Langloisia matthewsii, L. punctata, Leptodactylon californicum, Linanthus androsaceus, Loeselia ciliata, L. coerulea, L. glandulosa, L. mexicana, Navarettia breweri, N. intertexta, N. squarrosa, Phlox divaricata, P. drummondii, P. maculata, P. paniculata, P. stansburyi, Polemonium grandiflorum, P. leptans, P. mexicanum, P. pulcherrimum*.

POLYGALACEAE
18 genera; 950 species

With the exception of its absence in the western Pacific, the family can be considered cosmopolitan. Some members are cultivated ornamentals; several species of *Polygala* have been medicinally used.

Positive alkaloid tests have been recorded and small amounts of α-carbolines have been identified in *Polygala tennuifolia*. In this study, positive tests were also obtained for *Comesperma retuam*,

Monninia jalapensis (3/4), *Monninia sp.*, *Muraltia alopecuroides,*
M. heisteria, M. pauciflora, M. satureioides (1/2), *Polygala affinis,*
P. albidi, P. bracteolata, P. chinensis, P. ericaefolia, P. floribunda,
P. fruticosa, P. hottentotta (1/2), *P. kalaxariensis, P. lutea, P. lycopo-*
dioides, P. myrtifolia (1/3), *P. paniculata* (2/5), *P. pinifolia, Polyga-*
la spp. (2/13), *P. uncinata, P. virgata, Securidaca longipedunculata*
(1/2), *S. rivinaefolia.*

The following were negative: *Bredemeyera lucida, Comesperma
volubile, Diclidanthera sp., Epirixanthes cylindrica, Monninia
schlechtendaliana, M. sylvatica, Moutabea guianensis, Moutabea
sp., Muraltia divaricata, M. filiformis, M. macroceras, M. rham-
noides, Polygala angustifolia, P. arenicola, P. aspalathe, P. arillata,
P. brevifolia, P. celosioides, P. comata, P. galpinii, P. japonica,
P. lancifolia, P. lingustroides, P. macrodenia, P. moguiniana,
P. mollis, P. pauciflora, P. pringlei, P. senega, P. spectabilis,
P. sphenoptera, P. triphylla, Securidaca calophylla, S. diversifolia,
S. hostmanni.*

POLYGONACEAE
51 genera; 1,150 species

This family is almost cosmopolitan but is especially prominent in
the north temperate zone. Members are used as food, timber, tanning materials, and cultivated ornamentals.

Alkaloids have been detected in a few genera and some of these
have been characterized as protoalkaloids (e.g., benzylamines).

One hundred and eighty-two samples representing 122 species
were tested; the following were positive: *Antigonon flavescens,
Coccoloba floribunda* (2/3), *Coccoloba spp.* (2/2), *C. triplaris,
Gymnopodium floribundum, Muehlenbeckia sp.* (1/2), *Polygonum
limbatum, P. senegalense, P. serrulatum, Rumex lanceolatus.*

Negative tests were obtained with the following: *Antigonon lep-
topus, Brunnichia cirrhosa, Chorizanthe angustifolia, C. brevicor-
nu, C. corrugata, C. diffuse, C. membranacea, C. thurberi, C. uria-
ristata, Coccoloba ascendens, C. barbadensis, C. cozumelensis,
C. fallax, C. ilheensis, C. schiediana, Coccoloba spp.* (3), *C. uvif-
era, C. williamsii, Emex australis, Eriogonum angulosum, E. cine-
reum, E. deserticola, E. elongatum, E. heermanii, E. inflatum,*

E. kennedyi, E. latifolium, E. mohavense, E. molestum, E. nudum, E. parvifolium, E. umbellatum, Fagopyrum cymosum, Muehlenbeckia adpressa, M. axillaris, M. complexa, M. monticola, M. platyclada, M. tamnifolia, Oxygonum dregeanum, O. sinuatum, Oxygonum sp., Oxytheca dendroidea, Polygonella polygama, Polygonum acuminatum, P. amphibium, P. arenastrum, P. aviculare, P. caespitosum, P. chinense, P. coccinum, P. cristatum, P. cuspidatum, P. erectum, P. hydropiper, P. hydropiperoides, P. lapathifolium, P. mexicanum, P. nepalense, P. nodosum, P. orientale, P. pedunculare, P. pennsylvanicum, P. perfoliatum, P. persicaria, P. plebium, P. portoricense, P. pseudojaponicum, P. pterocarpum, P. pulchrum, P. punctatum, P. runcinatum, P. sagittatum, P. salicifolium, P. scandens, P. senticosum, Polygonum spp.* (4), *P. thunbergii, Pterococcus africanus, Pterostegia drymarioides, Rumex acetosella, R. angiocarpus, R. brownii, R. conglomeratus, R. cordatus, R. crassus, R. crispus, R. dentatus, R. hastatulus, R. japonicus, R. madaio, R. maritimus, R. obtusifolius, R. paucifolius, R. rhodesicus, R. sagittatus, Rumex spp.* (3), *R. woodii, Ruprechtia coriacea, R. fusca, R. laxiflora, R. pallida, Triplaris cumingiana, T. surinamensis, T. tomentosa.*

PONTEDERIACEAE
7 genera; 31 species

A family of tropical and warm areas, particularly of North America, the Pontederiaceae has a few north temperate species. Possibly the most familiar is the aquatic weed *Eichhornia.*

Eichhornia crassipes had been found earlier to contain putrescine, spermidine, and other polyamines; two out of four samples of it, along with *E. speciosa* and *Monochoria hastata* (1/2) gave positive tests in this study. The following were negative: *Heteranthera dubia, H. speciosa, H. reinfirmis, Monochoria vaginalis, Pontederia cordata, P. lanceolata.*

PORTULACACEAE
21 genera; 400 species

A few edible species are found in this family, which inhabits warm to tropical areas; others are used as ornamentals.

Alkaloids are known in but two genera of the family, including *Portulaca oleracea,* which gave one positive result in five samples tested in this study. Other positives included: *Lewisia rediviva* (1/2), *Montia perfoliata* (1/2), *Portulaca foliosa, P. pilosa* (1/2), *Talinum paniculatum* (1/2).

Species of several other genera were negative: *Anacampseros subnuta, A. telephiastrum, A. ustalata, Calandrinia balonensis, C. tuberosa, Calyptridium monandrum, C. umbellatum, Ceraria namaquensis, Claytonia virginica, Montia lamprosperma, M. mexicana, M. perfoliata, M. sibirica, Portulaca bicolor, P. coronata, P. cyanosperma, P. filifolia, P. hatschbachii, P. kermesiana, P. lanceolata, P. mucronata, P. mundula, Portulaca spp.* (3), *Portulacaria afra, Talinopsis frutescens, Talinum caffrum, T. crispatulum, T. cuncifolium, Talinum spp.* (2).

POTAMOGETONACEAE
2 genera; 90 species

This is a family of cosmopolitan fresh water herbs. Two species of *Potamogeton* and one of *Phyllospadix* have been reported to contain alkaloids. In this survey, *Potamogeton indicus* gave a positive test; others did not: *Potamogeton epihydrus, P. foliosus, P. gramineus, P. octandrus, P. pectinatus, P. perfoliatus, P. polyconus, P. polygamus, P. richardii, Zostera sp.*

PRIMULACEAE
22 genera; 800 species

A subcosmopolitan family of the northern hemisphere, the Primulaceae are known for their garden flowers. Occasional alkaloids have been noted in *Cyclamen* and *Primula.*

The following gave positive tests in this study: *Anagallis arvensis* (1/7), *Lysimachia ciliata, L. ruhmeriana, L. vulgaris* (1/3), *Primula alpicola, P. floribunda, P. sikkimensis.*

Negative species included the following: *Anagallis sp., Androdace umbellata, Ardisiandra wettsteinei, Dodecatheon jeffreyi,*

D. meadia, Lysimachia ardisioides, L. capillipes, L. quadrifolia, L. mauritana, L. recurvata, Primula auricula, P. candelabra, P. denticulata, P. japonica, Samolus ebracteatus, S. parviflorus, S. porosus, S. repens, S. valerandi, Trientalis latifolia.

PROTEACEAE
75 genera; 1,350 species

The Proteaceae occur mostly in the tropics and subtropics of the southern hemisphere, especially Australia and southern Africa.

Alkaloids, including pyrrolidines and tropanes, are known. This study indicated positive tests for the following: *Adenostephanus* (= *Euplassa*) *guianensis, Leucadendron sp., Persoonia toru, Protea laurifolia* (1/2), *P. micans, P. neriifolia.*

These species were negative: *Aulax cneorifolia, A. pallasia, Austromuellera trinervia, Banksia dentata, B. marginata, Beauprea sp., Brabejium stellatifolium, Conospermum mitchellii, C. patens, C. taxifolium, Euplassa legalis, Faurea macnaughtonii, F. saligna, F. speciosa, Finschia carrii, F. chloroxantha, F. rufa, Grevillea aquifolium, G. decora, G. huegelii, G. heterochroma, G. ilicifolia, G. papuana, G. parallela, G. robusta, G. rubiginosa, Grevillea spp.* (3), *G. wickhamii, Hakea acicularis, H. muelleriana, H. persiehana, H. rostrata, H. rugosa, H. suberea, Helicia albiflora, H. clemensiae, H. cochinchinensis, H. formosana, H. hypoglauca, H. insculpta, H. obtusata, Helicia sp., Knightsia diplanchii, K. excelsa, Leucadendron abscendens, L. argenteum, L. concinum, L. discolor, L. fusciflorum, L. grandiflorum, L. lanigerum, L. plumosum, L. spathulatum, L. strictum, L. tortum, L. venosum, Leucospermum album, L. attenuatum, L. candicans, L. catherinae, L. conocarpum, L. crinitum, L. lineare, L. nutans, L. prostratum, L. puberum, L. saxosum, Macadamia ternifolia, Mimetes lyrigera, Paranomus medius, P. reflexus, Paranomus sp., Persoonia falcata, P. juniperana, Protea acaulis, P. arbourea, P. caffra, P. cynaroides, P. eximia, P. gaugedi, P. glabra, P. humiflora, P. macrocephala, P. multibracteata, P. petiolaris, P. pulchra, P. repens, P. revoluta, P. rhodantha, Protea sp., P. subvestita, P. susannea, P. welwitschii, Rhapala rhombifolia, Roupala sp., Roupala comsimilis, R. montana, Roupala spp.* (5), *Serruria acrocarpa, S. adscendens, S. artemesiaefolia, S. bolusii,*

S. florida, S. fuscifolia, S. knightii, S. kraussii, Spatalla squamata, Stenocarpus salignus, Stenocarpus sp.

PUNICACEAE
1 genus; 2 species

When not cultivated elsewhere, this family is found from southeastern Europe to the Himalayas. It is noted for the pomegranate, *Punica granatum*, known since biblical times. This has long been known to be alkaloidal, and nine of 13 samples gave positive tests in this survey.

PYROLACEAE
4 genera; 42 species

The family ranges from the north temperate zone to Sumatra with some representatives in the south temperate zone. Nine samples, which included members of the four genera of the family, were negative for the presence of alkaloids: *Chimaphila umbellata, Monotropa hypopithis, M. uniflora, Pyrola rotundifolia, P. secunda, Sarcodes sanguinea.*

Cronquist has split the family, assigning *Monotropa* and *Sarcodes* to a separate family, Monotropaceae.

Q

QUIINACEAE
4 genera; 44 species

This is a family of tropical America, especially Amazonia.

Seven samples representing seven species were tested without positive results: *Quiina cruegeriana, Q. glaziovii, Q. leptoclada, Q. pteridophylla, Quiina sp., Q. tinifolia, Tourolia guianensis.*

R

RAFFLESIACEAE
8 genera; 50 species

In this tropical family with a few species in temperate zones, the most famous genus is *Rafflesia,* which produces the largest flower in the plant world.

The family is not known for alkaloids; Cronquist assigns *Mitrastemon* (= *Mitrastemma)* to its own family, Mitrastemmataceae, but admits, fortunately, that this is a matter of opinion and open to reasonable argument to the contrary.

A positive alkaloid test was obtained from an undetermined *Pilostyles sp.,* but another, *P. uleisolmis,* as well as *Mitrastemon yamamotoi,* gave negative results.

RANUNCULACEAE
58 genera; 1,750 species

Primarily a north temperate family, the Ranunculaceae is familiar to us as the family of buttercups and other weeds.

Aporphine and benzylisoquinoline alkaloids are found throughout the family, and several of the genera known to be positive were also found so in the survey reported here: *Aconitum ferox, Anemonella thalictroides, Aquilegia bertolonii, A. mckenna, Delphinium cardinale, D. glaucum* (2/2), *Thalictrum dasycarpum* (7/7), *T. fendleri, T. hernandezii, T. polygamum, Xanthorhiza apiifolia.*

In addition, the following were positive: *Anemone alpina, A. pulsatilla, A. sylvestris, Aquilegia flabellata, A. glandulosa, A. scopulorum, A. vulgaris, Clematis chinensis, C. paniculata, Clematopsis scabiosifolia* (1/3), *Colubrina reclinata, Delphinium parishii, D. parryi, D. pedatisectum, Delphinium sp.* (1/2), *Pulsatilla alpina, Ranunculus cantonensis, R. dichotomus, R. multifidus* (1/2), *R. weryii, Thalictrum caffrum* (2/2), *T. dioicum* (2/2), *T. rhynchocarpum, Thalictrum sp., Trollius albiflorus.*

These species were negative: *Actaea rubra, Adonis aestivalis, Anemone ritifolia, Anemone sp., Anemone virginiana, A. vitifolia, Aquilegia alpina, A. jonesii, Caltha howellii, C. palustris, Cimicifuga racemosa, Clematis aristata, C. brachiata, C. dioica, C. dioscoreifolia, C. drummondii, C. glycinoides, C. gouriana, C. henryi, C. leptophylla, C. leschenaultii, C. ligusticifolia, C. microphylla, C. oweniae, C. pitcheri, C. pycinoides, C. sericea, C. taiwaniana, C. virginiana, C. welwitschii, Clematopsis homblei, Coptis groenlandica, Glaucidium palmatum, Hepatica triloba, Knowltonia vesicatoria, Paeonia albiflora, P. brownii, P. californica, P. lactiflora, Ranunculus abortiva, R. aquatilis, R. bulbosus, R. delphinifolius, R. donianus, R. forreri, R. japonicus, R. lappaceus, R. macounii, R. pseudolowii, R. repens, R. sibbaldiaefolius, Ranunculus sp., R. stolonifer, Thalictrum minus, T. sessile.*

RESEDACEAE
6 genera; 75 species

This is a family of the Old World, especially of the temperate zone. It yields some dyestuffs and a few ornamentals.

Glucosinolates and an assortment of odd amino acids and alkaloidlike compounds are known, particularly in the genus *Reseda.*

Six samples were tested and gave one positive, previously known result, *Reseda luteola* (1/2), and three negatives: *Oligomeris oregeana, O. linifolia,* and *Reseda lutea.*

RESTIONACEAE
38 genera; 400 species

Confined to the southern hemisphere, the Restionaceae occur especially in Australia and South Africa. Some species are used locally as thatch.

Alkaloids are not known in the family. In this study, 27 samples representing 21 species gave but one positive test: *Willdenovia striata.*

The remainder tested were negative: *Cannomois virgata, Chondropetalum ebracteatum, C. mucronatum, C. paniculatum, Elegia galpinii, Elegia sp., E. verticillaris, Hypodiscus albo-aristatus,*

H. aristatus, Leptocarpus simplex, Restio cincinnatus, R. filiformis, R. perplexus, R. sieberi, Restio spp. (2), *Staberhoa cernua, Thamnochortus argenteus, T. dichotomus, T. fruticosus.*

RHABDODENDRACEAE
1 genus; 6 species

Found in the northern part of South America, this dimunitive family has no recorded alkaloids; a sample of *Rhabdodendron amazonicum* and *Rhabdodendron demidatus* gave a negative test for these substances.

RHAMNACEAE
53 genera; 875 species

The Rhamnaceae are cosmopolitan with concentration in tropical and warm regions. They furnish a well-known laxative Cascara (*Rhamnus purshianus*), as well as edible fruits, dyes, and ornamentals.

Alkaloids have been found in the family and several species known to be positive were also identified in this study: *Alphitonia macrocarpa, Ceanothus americanus* (2/2), *C. integerrimus* (1/2), *Hovenia dulcis* (1/3), *Zizyphus jujuba* (1/3), *Z. mauritania* (2/4), *Z. mucronata* (3/7), *Z. oenoplia.*

Other positive species included: *Adolphia infesta* (1/3), *Ampelozizyphus amazonicus* (1/2), *Ceanothus microphyllus, C. palmeri, C. spinosus* (2/2), *Colubrina reclinata, Condalia obtusifolia* (1/2), *Emmenospora alphitonoides, Gouania longipetala, G. lupuloides, G. polygama* (1/4), *Krugiodendron ferreum* (2/2), *Phylica dodii* (1/4), *P. olaefolia* (1/2), *P. paniculata, P. pubescens, P. rogersii* (2/2), *P. spicata, P. stipularis* (3/3), *P. rigidifolia, Rhamnus acuminatifolia* (1/6), *R. prunoides.*

Negative species included the following: *Alphitonia excelsa, A. neocaledonica, Berchemia floribunda, B. lineata, B. racemosa, Ceanothus coeruleus, C. cordulatus, C. crassifolius, C. cuneatus, C. divaricatus, C. greggii, C. incanus, C. megacarpus, C. pinetorum, Ceanothus sp., C. tomentosus, C. velutinus, C. verrucosus, Colubrina asiatica, C. ferruginosa, C. glomerata, C. greggii, C. palmeri, C. reclinata, Condalia brandegei, C. fasciculata,*

C. lycioides, C. mexicana, C. parryi, C. ondalia sp. aff. *warnockii, Crumenaria polygaloides, Gouania longispicata, G. polygama, Gouania spp.* (2), *G. stipularis, G. tomentosa, Karwinskia humboldtiana, K. mexicana, K. pubescens, Microrhamnus ericoides, Phylica cryptandroides, P. ericoides, P. oxillaris, Phylica spp.* (2), *P. villosa, Phyllogeiton discolor, P. zeyheri, Pomaderris kumeraho, P. oraria, P. phylicifolia, Rhamnidium sp., Rhamnus californica, R. caroliniana, R. formosana, R. ilicifolia, R. leptophyllus, R. microphylla, R. nakaharai, R. oenoplia, R. pianensis, R. purshianus, R. sectipetala, R. serrata, Rhamnus spp.* (2), *Sageretia* aff. *elegans, Scutia buxifolia, S. myrtina, Smythea lanceata, Spyridium parvifolium, S. vexilliferum, Stenanthemum scortechnii, Ventilago ecorallata, V. maderaspatana, Ventilago sp.* cf. *microcarpa, Zizyphus abyssinica, Z. amole, Z mexicana, Z. xylophorus.*

RHIZOPHORACEAE
16 genera; 130 species

A family primarily of the Old World tropics, the Rhizophoraceae are used in some areas for timber and tanning.

Tropanes and thiolane-type alkaloid substances have been found in the family. Tests here included 35 samples, of which five species gave positive tests: *Brugiera sexangula, Carallia brachiata, C. latifolia* (these three were previously known), *Brugiera exaristata* (1/2), *Cassipourea gerrardi* (3/3).

Other species of the same genera were negative: *Brugiera conjugata, B. gymnorhiza, B. parviflora, B. rheedii, Carallia integerrima, Cassipourea gummiflua, C. axillaris, Rhizophora apiculata, R. mangle, R. mucronata, R. samoensis, R. stylosa.*

ROSACEAE
107 genera; 3,100 species

This is a well-known, subcosmopolitan family especially of warm temperate regions. Many of our common fruits, fragrant oils, and popular garden plants are found in this family.

Alkaloids occasionally occur (diterpenes, hydroxytryptamine, phenylalklamines) and cyanogenesis is common in many seeds of the fruits. Nonetheless, such compounds are not that common considering the size of the family.

In the survey reported here, positive tests were given by the following: *Aronia arbutifolia* (2/2), *A. atropurpurea, A. melanocarpa, Dryas suendermanni, Geum pedersii, Hirtella americana* (1/2), *Licania sp.* (1/3), *Osmaronia* (= *Oemleria*) *cerasiformis, Parinari capensis, P. glandulosa, Potentilla haematochrus, Prunus selowii, Prunus sp.* (1/5), *Purshia glandulosa* (2/2), *Pyracantha lelandii, Rhodotypos scandens* (2/2), *Rubus odoratus, Spiraea japonica.*

Most of the species tested were negative: *Acaena elongata, A. eupatoria, A. microphylla, A. ovina, Aceana sp., Adenostoma fasciculatum, A. sparsifolium, Agrimonia eupatoria, A. hirsuta, A. parviflora, A. prolifera, Alchemilla elongata, A. procumbens, Amelanchier asiatica, A. canadensis, A. denticulata, A. humilis, A. pumila, Amelanchus oblongifolia, Amygdalus* (= *Prunus*) *persica, Argentina larrea, Aruncus dioicus, Cercocarpus betuloides, C. lediflorus, C. macrophyllus, C. paucidentatus, Chaenomeles legenaria, Chamaebatia foliolosa, Chamaevatiaria millefolium, Chrysobalanus icaco, Cliffortia baccans, C. burchellii, C. crenata, C. cuneata, C. falcata, C. graminea, C. ilicifolia, C. nitidula, C. odorata, C. polygonifolia, C. ruscifolia, C. strobelifera, Coleogyne ramosissima, Cotoneaster acuminata, C. bullata, C. congesta, C. dammeri, C. francheti, C. morrisonensis, C. salicifolia, Couepia bracteosa, C. divaricata, C. leptostychya, C. polyandra, Cowania ericaefolia, C. mexicana, C. plicata, Crataegus columbiana, C. mexicana, C. michauxii, C. monogyna, C. phaenopyrum, C. pinnatifida, C. pyracanthoides, C. rosei, Crataegus sp., Cydonia japonica, Dyras drummondii, D. octopetala, Duchesnea indica, Eriobotrya japonica, Exochorda racemosa, E. serrayifolia, Fragaria collina, F. mexicana, F. platypetala, Fragaria spp.* (3), *F. vesca, Galium asperellum, Galium sp., Geum borisii, G. japonicum, G. pyrenaicum, G. virginicum, Grielum humifusum, Heteromeles arbutifolia, Hirtella aff. americana, H. druidsii, H. hebeclada, H. paniculata, H. proealta, H. racemosa, Hirtella spp.* (2), *H. tentaculata, H. triandra, Holodiscus discolor, H. dumosus, Horkelia cuneata, Ivesia santolinoides, Leucosidea sericea, Licania arborea, L. biglandulosa, L. canescens, L. glabra, L. heteromorpha, L. lon-*

gistyla, L. micrantha, L. octandra, L. rigida, Licania sp. aff. *sprucei, L. vaupesana, Lindleyella mespiloides, Malus sieboldii arborescens, Mespilus germanica, Mitchella repens, Moquilea sp., M. utilis, Neillia sinescens, N. thyrsifolia, Neonatea sp., Parinari curatellifolia, P. nonda, P. papuanum, P. glaberrimum, P. insularum, P. nonda, P. laurinum, Photinia arbutifolia, P. glabra, P. parviflora, P. serrulata, P. taiwanensis, P. villosa, Physocarpus capitatus, P. opulifolius, Potentilla candicans, P. egedei, P. foesteriana, P. fruticosa, P. fulgens, P. gracilis, P. megaleasa, P. norvegica, P. papuana, P. petinisecta, P. pumila, P. recta, P. richardii, P. simplex, Poterium polyamum, Primarium obtusifolium, Prinsepia scandens, Prunus americana, P. andersonii, P. avium, P. buergeriana, P. brasiliensis, P. brachybotria, P. capuli, P. caroliniana, P. cortapico, Prunus sp.* cf. *costata, P. fasciculata, P. fremontii, P. gazelle-peninsulae, P. glomerata, P. grisea, P. ilicifolia, P. lyonii, P. oligantha, P. pullei, P. schlerchteri, P. serrulata, P. jamasakura, P. japonica, P. mume, P. ochoterenae, P. persica, P. phaeosticta, P. sargentii, P. serotina, P. serotina virens, P. umbellata, P. virginica, P. yedoensis, Pseudocydonia sinensis, Purshia tridentata, Pyracantha atlantoides, P. koidzumi, Pyrus americana, P. pyrifolia, P. sikkimensis, Rhaphiolepis liukiuensis, R. umbellata, Rosa bracteata, R. carolina, R. gentifolia, R. gymnocarpa, R. laevigata, R. rosifolia, R. sambrica pubescens, R. taiwanensis, R. webbiana, R. wichusaiana, Rubus adenotrichus, R. alnifoliatus, R. archboldianus, R. buergeri, R. calycinoides, R. cissoides, R. conduplicatus, R. elegans, R. ellipticus, R. euphlebophyllus, R. fasciculatus, R. fraxinifolius, R. lacinato-stipulatus, R. laettervirdis, R. lambetianus, R. leucodermis, R. lorentzianus, R. ludwigii, R. moluccanus, R. montiswilhelmi, R. parifolius, R. parviflorus, R. pinnatus, R. piptopetalus, R. pumilus, R. reflexus, R. rigidus, R. rosaefolius, R. semialata, R. sinkoensis, Rubus spp.* (5), *R. taiwanianus, R. tiliaceus, R. triloba, R. trivialis, R. ursinus, R. urticifolius, R. vitifolius, Sanguisorba minor, S. occidentalis, Sericotheca fissa, S. pachydisca, S. veltulina, Spiraea japonica formosana, S. prunifolia, S. tomentosa, S. vaccinifolia, S. veitchii, Stephanandra incisa, Stranvaesia davidiana, Vauquelinia angustifolia.*

Several of the genera listed here as Rosaceae have since been placed in a separate family, the Chrysobalanaceae. These include

Hirtella, *Licania*, *Parinari*, *Chrysobalanus*, *Couepia*, *Moquilea* (= *Licania*), *Parinarium* (= *Parinari*).

RUBIACEAE
630 genera; 10,000 species

This is another of the very large families of the plant kingdom, pantropical and subtropical in distribution with some members extending into the temperate zones of both the northern and southern hemispheres. It is of great economic importance as the source of coffee, several drugs, native medicines, dyes, and many ornamentals.

Several genera of the family have been assigned to a separate family, Naucleaceae, by some taxonomists. The family is considered here in its extended and generally accepted sense. Many of the tests on the relatively uncommon species were conducted on small samples gleaned from herbarium specimens.

The family is known for alkaloids of many structural types, including those of quinine, emetine, strychnine, and harman. Understandably, many of the species tested in this survey of over 1,100 samples were found to be alkaloid-positive, having been recorded as such in earlier literature: *Anthocephalus cadamba* (1/2), *Antirhea putaminosa* (2/2), *Bobea elatior*, *Borreria verticillata*, *Canthium odoratum*, *Cephalanthus occidentalis* (4/8), *C. pubescens*, *Exostema sancta-luciae*, *Gardenia jasminoides* (2/3), *Hamelia patens* (7/10), *Hodgkinsonia ovatiflora*, *Isertia hypoleuca* (2/2), *Mitragyna africana*, *M. ciliata* (3/3), *M. inermis*, *M. javanica*, *M. rotundifolia* (2/2), *M. rubrostipulata* (2/2), *M. speciosa* (164 samples of this species were tested in conjunction with the collection of large quantities for extraction and isolation of specific alkaloids), *M. stipulosa* (11/11), *Nauclea latifolia*, *N. maingayi*, *M. officinalis*, *Oldenlandia biflora*, *Paederia rigida* (1/2), *Pinkneya pubens* (2/2), *Rubia cordifolia* (1/2), *Sarcocephalus esculentus*, *Uncaria bernaysii*, *U. ferrea* (1/3), *U. longiflora*.

Alkaloids of the following species that tested positive in this study have apparently not yet been reported: *Acranthera velutinervia*, *Alseis floribunda*, *Alseis sp.*, *Antirhea tenuifolia*, *Asperula cynanchica*, *Basanacantha spinosa*, *Bathysa australis*, *B. meridionalis*, *Borreria corymbosa* (2/2), *B. dibrachiata*, *B. saxicola*, *B. scabra*

(1/3), *Borreria sp.* (4/4), *B. suaveolens, B. valerianoides, Canthium hispidum, C. horridum, C. huillense, C. randii, Carpacoce scarpa, Catesbaea spinosa, Cephaelis barcellona* (leaf, stem, root), *Cephaelis spp.* (2), *C. tomentosa* (3/5), *Cephalanthus amygdalifolius, C. glabratus, Chimarrhis turbinata* (4/4), *Chiococca alba* (1/8), *Coutarea hexandra* (3/5), *Coutarea sp.* (1/2), *Crusea spp.* (2/3), *Diodia natalensis, D. prostrata, D. virginiana, Diplospora sp.*, *Didrichletia* (= *Carphalea*) *pubescens, Enterospermum* (= *Tarenna*) *rhodesiacum* (1/2), *Exostema caribaeum* (2/5), *E. longifolium, Faramea spp.* (5/15), *Feretia aeruginescens, Galium bussei* (1/2), *G. capense, G. tomentosum, Gardenia macgilliravei, G. spathulifolia* (1/3), *Genipa brasiliensis* (1/2), *G. clusifolia, Hamelia erecta, Hamelia sp.* (1/3), *H. versicolor* (3/3), *Hedyotis sp., H. tenelliflora, Hillia parasitica* (1/3), *Isertia kaeniana* (2/3), *I. longifolia* (3/3), *I. rosea* (1/3), *Ixora beckleri, Ixora sp.* (1/2), *I. undulata, I. williamsii* (2/3), *Kohautia amatymbica* (1/2), *Kohautia sp.* (1/2), *Kutchubea sp., Leptactina benguelensis* (2/2), *Mitragyna parviflora* (1/2), *Morinda sp.* (1/2), *Mussaenda erythrophylla* (1/3), *Myrmecodia antionii, Nauclea rhynchophylla, Naculea sp., Oldenlandia affinis* (1/2), *O. auricularia, O. pellucida, Paederia foetens, P. scandens* (1/2), *Palicourea condensata, P. corymbifera* (5/5), *P. crocea* (2/2), *P. galleotiana* (1/2), *P. guianensis, P. macrophylla, P. obscurata, P. rigida, Palicourea spp.* (5/13), *Pavetta assimilis* (1/3), *P. barbertonensis, P. edentula, P. eyelsii, P. gracillima, P. harborii, P. lanceolata* (2/2), *P. macrophylla, P. revoluta, P. schumanniana, P. zeyheri, Pentas angustifolia, P. nobilis, Portlandia grandiflora* (5/5), *Posoqueria trinitatis, Pseudocinchona* (= *Pausynstalia*) *africana* (5/5), *Psychotria barbiflora* (2/2), *P. broweri* (2/2), *P. coelospermum, P. horizontalis, P. lupulina, P. micodon, P. oleoides, P. orinoides* (root) (1/2), *P. ovoidea, P. polycephala, P. rubra* (1/3), *Psychotria spp.* (11/22), *P. suturella* (1/2) (leaf and fruit), *Randia* aff. *cinerea, R. cochinchinensis* (1/2), *R. formosa* (1/2), *Relbunium sp.* (1/2), *Remijia amazonica, R. peruviana, Remijia sp., Rondeletia galeottii, Rothmannia capensis, Sarcocephalus esculentus, Sickingia* (= *Simira*) *sp.* (stems) (1/2), *S. tinctoria* (2/2), *Standleya pinnata, Straussia* (= *Psychotria*) *kaduana, Tarenna attenuata, Thysanospermum* (= *Coptosapelta*) *diffusum* (1/3), *Tricalysia cacondensis, Uncaria*

guianensis (1/2), *U. macrophylla* (2/2), *U. philippensis, U. sinensis, U. tomentosa* (3/3), *Vangueria madagascariensis.*
The following species were negative: *Acranthera frutescens, Adina cordifolia, A. galpini, A. globiflora, A. microcephala, A. multifolia, A. pilulifera, A. polycephala, A. racemosa, A. rubella, A. rubescens, Adina sp., Agathisanthemum* (= *Oldenlandia*) *bojeri, A. globosum, Alibertia colulii, A. edulis, A. grandiflora, A. magna, A. sessilis, Amaioua corymbosa, A. guianensis, Amaracarpus spp.* (2), *Anthocephalus* (= *Oldenlandia*) *indicus, A. macrophyllus, Argostemma bryophilum, Atherospermum aethiopicum, A. herbaceum, A. littoreum, A. panuculatum, A. randii, Atherospermum sp., Badusa corymbifera, Basanacantha phyllosepala, Basanacantha spp.* (4), *B. spinosa, Bathysa stipularis, Blepharidium mexicanum, Borreria arvensis, B. capitata, B. eryngioides, B. laevis, B. latifolia, B. ocymoides, B. poaya, B. scabra, B. stricta, B. suaveolens, B. subvulgata, Bouvardia chrysantha, B. leiantha, B. linearis, B. longiflora, B. multiflora, B. scabida, B. ternifolia, B. viminalis, Burchellia bulbalina, B. capensis, Calycophyllum candidissimum, C. spruceanum, Canthium coprosmoides, C. frangula, C. gilfillanii, C. grenzii, C. inerme, C. lactescens, C. latifolium, C. murrillii, C. queinzii, C. setiflorum, C. vulgare, Capirona decorticans, Carinta* (= *Geophila*) *sp., Cephaelis barcellana, C. colorata, C. elata, C. mucosa, Cephaelis spp.* (3), *Cephalanthus berlandieri, C. natalensis, C. salicifolius, Chasalia carviflora, Chiococca pachyphylla, Chomelia obtusa, C. cordifolia, Cinchona sp., Coccocypselum lanceolatum, C. repens, Coccocypselum spp.* (2), *Coffea arabica, C. bengalensis, Coffea spp.* (2), *Conostomium natalense, Coprosma acerosa, C. arborea, C. australis, C. chessemanii, C. cunninghamii, C. foetidissima, C. foliosa, C. microcarpa, C. nadeaudiana, C. novoguinensis, C. parviflora, C. pilosa, C. prisca, C. pseudocuneata, C. repens, C. rhamnoides, C. robusta, Coprosma spp.* (3), *C. tenuifolia, Coussarea hydrangeifolia, C. pumiulata, Coutarea latifolia, C. pterospermia, C. speciosa, Coutarea sp., Craterispermum laurinum, Crossopteryx febrifugum, Crusea alloca, C. brachyphylla, C. calocephala, C. coccinea, C. coronata, C. hispida, Crusea sp., C. subulata, Damnacanthus indicus, Declieuxia dusenii, Declieuxia spp.* (8), *D. spergulifolia, Deppea grandiflora, Didymaea mexicana, Didymaea sp., Diodia arenosa, D. brasiliensis, D. hispidula, Diodia spp.* (5), *D. teres,*

Diplospora australis, Duggena hirsuta, Duroia genipoides, D. hirsuta, Elaeagia sp., Emmenopterys henryi, Emmeorhyza umbellata, Ernodea angusta, Exostema brachycarpum, E. ilegana, E. lineatum, E. mexicanum, E. peruvianum, Fagodia agrestis, F. fragrans, F. monticola, F. odorata, F. tetraquetra, Faramea anisocalyx, F. capillipes, F. maynensis, F. occidentalis, F. quinqueflorum, F. rectinervia, Faramea spp. (2), *F. subbasiliaris, Ferdinandusa elliptica, F. paraensis, F.* aff. *paraensis, F. rudgeoides, Feretia canthoides, Galium angustifolium, G. aparine, G. aschenbornii, G. formosana, G. hallii, G. mexicanum, G. mollugo, G. nuttallii, G. pubens, Galium spp.* (3), *G. spurium, G. stellatum, G. triflorum, Galopina aspera, G. circaeoides, Gardenia angusta, G. angustifolia, G. brachythamnus, G. cornuta, G. florida, G. gummifera, G. jasminoides, G. latifolia, G. neuberia, G. resiniflua, Gardenia spp.* (2), *G. urvillei, Genipa americana, Genipa sp.* (2), *Gouldia terminalis, Grumilea* (= *Mapouria* = *Psychotria*) *capensis, G. kirkii, Guettarda combusii, G. crispifolia, G. platyphylla, G. scabra, Guettarda spp.* (7), *G. speciosa, G. uruguayensis, Hamelia rovirosae, Hamelia spp.* (2), *Hamiltonia suaveolens, Hedyotis acutangula, H. corymbosa, H. loganioides, Hedyotis sp., H. tenelliflora, Heinsia crinita, Hemidiodia sp., Hoffmannia chiapensis, H. cryptoneura, H. discolor, H. lenticillata, H. mexicana, Houstonia acerosa, H. angustifolia, Houstonia coerula, Hymenodictyon excelsum, H. floribundum, Hypnophyton formicarum, Isertia commutata, I. hoehnei, I. parviflora, Ixora beckleri, I. brachiata, I. coccinea, I. finlaysoniana, I. francavillana, I. macrothyrosa, I. odorata, I. pubescens, Ixora spp.* (3), *I. triflora, I. venulosa, Kohautia gracilifolia, Kohautia sp., Kraussia floribunda, Lagynias australis, Lasianthus chinensis, L. curtisii, L. plagiophyllus, L. tashiroi, L. teiheizanensis, Lipostoma capitatum, Luculia gratissima, L. intermedia, L. pinceana, L. yunnanensis, Machaonia brasiliensis, M. coulteri, Malanea macrophylla, M. sarmentosa, Manettia cordifolia, M. glaziowii, M. gracilis, M. quinquinervia, Manettia spp.* (2), *Mephitida* (= *Lasianthus*) *formosana, N. nigricarpa, Mitchella sp., Mitracarpum hirsutum, Morinda citrifolia, M. billiardieri, M. royoc, Morinda sp., M. umbellata, Mussaenda arcuta, M. frondosa, M. parviflora, M. pubescens, M. taihokuensis, Myrmeconauclea strigosa, Nauclea chalmersii, N. cordata, N. diderichii, N. esculenta, N. junghuhnii, N. orientalis, N. robinsonii, Nauclea spp.* (4), *N. subdita, N. tennui-*

flora, N. trillasii, N. undulata, Neonauclea angustifolia, N. bartlingii,
N. bernardei, N. chalmersii, N. fosteri, N. griffithii, N. gracilis,
N. oligophlebia, N. pallida, N. papuana, N. purpurescens, N. purpu-
rea, N. schlechteri, Neonauclea sp., Normandia neocaledonica, Nor-
mandia sp., Oldenlandia corymbosa, O. hedyotidea, O. herbacea,
O. pringlei, Opercularia diphylla, O. turpis, Ophiorrhiza dimor-
phantha, O. inflata, Otiophora calycophyllum, O. inyangana,
O. scabra, Oxyanthus gerrardii, Pachystigma sp., Paederia sp., Pa-
gamea duckei, P. guianensis, Pagamea spp. (4), *Palicourea barbin-*
ervis, P. fastigata, P. guianensis, P. paraense, P. platypodina,
P. sellowiana, Palicourea spp. (7), *Pavetta catophylla, P. cooperi,*
P. gestneri, P. gracilifolia, Pavetta sp., Pentanisia prunelloides,
P. schweinfurthii, Pentas purpurea, Plectronia parviflora, Plectro-
niella aemata, Posoqueria latifolia, Posoqueria sp., Psychotria ca-
pensis, P. cuspidata, P. daphnoides, P. erythrocarpa, P. hancornifo-
lia, P. hebochada, P. hexandra, P. inundata, P. involucrata,
P. laxissima, P. longipes, P. nervosa, P. nuda, P. oerstediana, P. papant-
lensis, P. patens, P. pinularis, P. pubescens, P. repens, P. serpens,
P. sessilis, P. sessilifolia, Psychotria spp. (12), *P. stachyoides,*
P. trinitensis, P. undata, Psyllocarpus laricoides, Pygmaeothamnus
zeyheri, Randia aculeata, Randia aff. *R. densiflora, R. dumetorum,*
R. lactovirens, R. longiflora, R. longispina, R. maculata, R. mitis,
R. tetracantha, R. ulaginosa, R. watsonii, Relbunium vile, R. equise-
toides, Remijia ferruginea, Retiniphyllum spp. (2), *Rhabdostigma*
(= *Kraussia*) *sp., Richardia brasiliensis, R. humistrata, Richardia*
sp., Rondeletia aff. *ligustroides, R. amoena, R. budelleioides,*
R. leucophylla, R. odorata, Rothmannia globosa, Rubia akane, Rud-
gea freemani, R. gardenioides, R. jasminoides, Rudgea spp. (2),
R. viburnoides, R. villosa, Sabicea cana, S. entebensis, S. trinitensis,
Salzmania nitens, Schradera sp., Scyphiphora hydrophylacea, Sick-
ingia (= *Simira*) *oliveri, Sipanea hispida, Spermacoce podocephala,*
S. stricta, S. tenella, Straussia (= *Psychotria*) *mariniana, Tapiphyl-*
lum sp., T. velutinum, Tarenna lancifolia, T. neurophylla, T. sambuxi-
na, Temnocalyx obovatus, Thieleodoxa sorbilis, Timonius affinis, Ti-
monius sp., T. timon, Tricalysia allenii, T. angolensis, T. lanceolata,
T. sonderiana, Tricalysia sp., Ucriana (= *Tocoyena*) *longifolia, Un-*
caria africana, U. attenuata, U. canescens, U. cordata, U. dasyneu-
ra, U. gambir, U. glabrata, U. hirsuta, U. ovalifolia, U. perrotteri,

U. rhyncophyllum, U. scandens, U. sclerophylla, U. sessilifructus, U. talbotii, U. tonkinensis, Uragoga (= Cephaelis) sp., Vangueria chartacea, V. infusta, V. longicalyx, V. randii, Vangueriopsis lancifolia, Warscewiczia coccinea, Xeromorphia sp., Xeromphia (= Randia) obovata, X. rudis, Zygoon (= Tarenna) graveolens.

RUTACEAE
161 genera; 1,700 species

This cosmopolitan, but mainly tropical, family furnishes our citrus fruits, many aromatic oils, and some medicinals, timber, and ornamentals.

Alkaloids are common throughout the family. The following species, previously reported to be alkaloidal, were confirmed so in the study described here: *Acronychia imperforata, A. pauciflora* (1/2), *Calodendrum capensis, Casimiroa edulis* (2/2), *Chloroxylon swietenia, Choisia ternata, Citrus grandis* (1/2), *Clausena brevistyla* (1/2), *C. lancium, Dictyoloma incanescens, Evodia alata* (2/2), *E. hortensis* (4/5), *E. meliaefolia, Fagara xanthoxyloides* (4/4), *Geijera salicifolia* (1/3), *Hortia arborea, Lunasia amara, Monniera* (= *Ertala*) *trifolia, Murraya paniculata* (2/4), *Phebalium nudum, Phellodendron amurense, Pilocarpus jaborandi, Ptelea trifoliata, Ravenia spectabilis, Ruta graveolens* (2/2), *Skimmia reevesiana* (1/4), *Toddalia asiatica* (1/2), *Zanthoxylum alatum* (1/2), *Z. americanum, Z. fagara* (1/2), *Z. nitidum, Z. oxyphyllum, Ziera smithii.*

Other positive species included the following: *Acronychia melicopoides, A. murina, A. pedunculata, A. suberosa, Adenandra uniflora, Agathosma betulina, A. capensis, A. crenulata, A. dielsiana, A. mucronulata, A. virgata, Amaryis elemifera* aff. *rekoi, A. sylvatica, Atalantia buxifolia* (2/2), *Barosma* (= *Angostura*) *sp., Diosma ramosissima, Erythrochiton brasiliensis, Esenbeckia berlandieri, Esenbeckia sp., Euxilophora paraensis* (1/2), *Evodia danielei, E. elleryana* (2/2), *E. hirsutifolia* (3/3), *Fagara davyii, F. humilis* (2/2), *F. megalismontana, F. martinensis* (3/3), *F. rhoifolia* (1/5), *Fagara sp.* (1/2), *F. thorncroftii, Flindersia chrysantha, Glycosmis cochinchinensis* (1/3), *Halfordia papuana, Limonia crenulata, Melicope perspecuinervia, M. stipitata, Melicope spp.* (3/4), *Micromelum minutum* (2/3), *Monathrocitrus cornuta, Murraya exotica* (2/2),

M. omphalocarpa, Murraya sp., Orixa japonica, Phellodendron chinense, Pilocarpus giganteus, P. longipes, Pilocarpus sp. (1/2), *Skimmia ariesanensis* (2/3), *Swinglea glutinosa, Teclea natalensis, Teclea sp., Toddalia aculeata* (stem) (1/2), *T. grandifolia, Triphasia trifolia, Vepris reflexa* (2/2), *Vepris sp.* (1/2), *V. undulata* (2/2), *Zanthoxylum avicennae* (3/3), *Z. blackburnia, Z. conspersipunctatum* (2/2), *Z. dominianum, Z. juglandifolium, Z. pluviatale, Z. pteropodium* (2/2), *Z. simulans.*

In spite of the abundance of alkaloids in the family, a number of species tested negative for these compounds: *Acmadenia densiflora, A. juniperiana, Acronychia cooperi, A. laurifolia, Achronychia* aff. *parryi, A. oblongifolia, Adenandra brachyphylla, A. cuspidata, A. fragrans, A. serpyllacea, Adiscanthus sp., Aegele marmelos, Agathosma bifida, A. cerefolium, A. collina, A. marifolia, A. minuta, A. ovata, A. peglerae, A. serpyllacea, Agathosma sp., Asterolasia muelleri, Boenninghausia albiflora, Boronia caerulescens, B. nana, B. parviflora, B. pilosa, Bosistoa sapindiformis, Casimiroa pubescens, Chalas exotica, C. paniculata, Citrus colocynthus, C. decumana, C. sunki, Citrus spp.* (4), *Clausena brevistyla, C. indica, Clausena sp., Coleonema album, Correa reflexa, Cusparia pentogyne, Decatropis bicolor, Dictyoloma incanescens, Diosma aspalathoides, D. vulgaris, Eremocitrus glauca, Eriostemon difformis, Esenbeckia intermedia, E. pumila, Evodia bonwickii, E. glauca, E. henryi, E. lepta, E. ptelaefolia, Evoida spp.* (4), *Evodiella hooglandii, Fagara capensis, F. hymenalis, F. kleinii, F. microcarpa, F. nitida, Feronia elephantum, Geijera salicifolia, Glycosmis citrifolia, G. pentaphylla, Hortia brasiliana, H. longifolia, Hortia spp.* (3), *Limonia crenulata, Melicope sessiliflora, M. stipitata, Metrodorea nigra, M. pubescens, Micromelum pubescens, Murraya koenigii, Myrtopsis spp.* (2), *Paramigya griffithii, Pelea elusisefolia, P. wawreana, Phebalium decamanum, P. nottii, P. rotundifolium, Phebalium sp., Phellodendron japonicum, Pilocarpus sp., Polyaster boronoides, Poncirus trifoliata, Raputia sp., Severina buxifolia, Thamnosma africanum, Zanthoxylum sp., Ziera compacta, Z. laxiflora.*

Evodia (= *Euodia*) is considered by some authors to be *Melicope* + *Tetradium.*

S

SABIACEAE
3 genera; 48 species

The family consists of trees, shrubs, and lianas of Southeast Asia and tropical America.

An early report of a positive alkaloid test in a species of *Meliosma* exists, but cyanogenesis is more characteristic of this small family. In the present instance, six species of *Meliosma* and one of *Sabia* failed to give positive tests: *Meliosma dentata, M. humilis, M. myriantha, M. rhoifolia, M. rigida, Meliosma sp., Sabia pauciflora*.

SALICACEAE
2 genera; 435 species

This is a subcosmopolitan family with concentration in the northern hemisphere. The family of the weeping willow, it has been known since biblical times and was the source of salicylic acid for the Greek physicians, which led, eventually, to the synthesis of aspirin for the relief of minor pain. Pliable, thin branches are used in basket weaving, and some timber is produced.

A couple of isolated reports of the presence of alkaloids in *Salix* and *Populus* have been recorded, but it is not known to be an alkaloidal family. In this study, positive tests were obtained for *Populus fremontii, Salix gracilistyla,* and *S. purpurea*. Eight other species of *Populus* and 24 of *Salix* were negative: *Populus alba, P. balsamifera, P. deltoides, P. grandidentata, P. monticola, P. tremuloides, P. trichocarpa, P. yunnanensis, Salix babylonica, S. bonplandiana, S. capensis, S. caprea, S. caroliniana, S. chinensis, S. cordata, S. discolor, S. fragilis, S. fulvopubescens, S. humilis, S. interior, S. laevigata, S. lasiolepis, S. lucida, S. nigra, S. oxylepis, S. rigida, S. scouleriana, S. sericea, Salix sp., S. subserrata, S. tristis, S. warburgii, S. woodii*.

SALVADORACEAE
3 genera; 11 species

The family is characteristic of the xerophytic areas of warm parts of the Old World. Twigs–"chew sticks"–are used by some African peoples as toothbrushes.

Five samples representing four species were tested for alkaloids with the following results: *Azima tetracantha* (1/3) was known to be positive while three species of *Salvadora* gave negative results: *S. angustifolia, S. indica, S. australis.*

SANTALACEAE
36 genera; 500 species

Members of the family are found in warm to tropical areas around the world. Essential oils (e.g., sandalwood), some edible fruits and tubers, as well as timber are obtained from the family.

Several genera/species of the family are known to contain alkaloids. In this study, positive results were obtained on testing *Exocarpos aphylla, Rhoiacarpos capensis, Thesium australe* (1/2), *T. hystrix* (1/2), *T. lanciulatum, T. magalismontanum,* and *T. virgatum.*

Alkaloid-negative species included the following: *Anthobolus filifolius, Colpoön compressum, Exocarpos homaloclada, E. neocaledonicus, E. phyllanthoides, Exocarpus sp., E. strictus, Henslowia frutescens, H. queenslandica, Osyridocarpos natalensis, O. compressa, Pyrolaria pubera, Santalum acuminatum, S. freycinetisum, S. lanceolatum, S. murrayanum, Scleropyrum auriantiacum, Thesium carinatum, T. eldule, T. exile, T. foliosum, T. hispidulum, T. juncifolium, T. namaquense, T. scabrum, Thesium spp.* (2), *T. strictum, T. subaphyllum, T. transvaalense.*

SAPINDACEAE
144 genera; 1,325 species

A few of the members of this family are from temperate regions, but most are found in warm to tropical areas. The family yields several edible fruits (e.g., akee, litchi), timbers, soap substitutes, and fish poisons (saponins). In South America, a caffeine-containing drink, guarana, is popular, and the western Amazonian Indians derive a stimulant drink from *Paullinia yoco*.

Although the xanthines (e.g., caffeine) are found in some members of the family and do not give demonstrable Dragendorff tests, there are other reports of unidentified alkaloids. In this survey, the previously known alkaloid-positive species *Cardiospermum halicababum* (1/7) and *Sapindus mukarossii* (1/5) were positive as were the following: *Atalaya hemiglauca, Dictyoneura bamberi, Dodonea viscosa* (1/10), *Neopringlea integrifolia, Serjania sp., Urvillea ulmaceae* (1/3), *Xanthoceras sorbifolia* (2/2).

Negative tests were obtained with the following: *Alectryon excelsum, Allophylus africanus, A. cobbe, A. divaricatus, A. edulis, A. gilioides, A. melanocarpus, A. natalensis, A. petiolulatus, A. serratus, Allophylus spp.* (2), *Antidesma parvifolium, Arytera sordida, Atalaya alata, A. varifolia, A. virens, Blighia sapida, B. unijugata, Cardiospermum corindum, Cardiospermum sp., Cupania glabra, C. oblongifolia, C. rubiginosa, C. vernalis, Cupanopsis macropetala, Dictyoneura obtusa, Dimocarpus longan, Distichostemon hispidulus, Dodonea attenuata, D. jamaicensis, D. stenophylla, Erioglossum rubiginosum, Erythrophysa transvaalensis, Euphoria longana, Filicium decidens, Guida acutifolia, G. coriacea, Harpullia cupanoides, H. imbricata, H. pendula, Heterodendron olaefolium, Hippobromus pauciflorus, Jagera pseudorhus, Koelrueteria formosana, K. paniculata, Lecaniodiscus fraxinifolia, Litchi chinensis, Matayba arborescens, M. elaeagnoides, M. guianensis, M. juglandifolia, Matayba sp., Melicocca bijuga, Mischocarpus retusus, Neopringlea viscosa, Pappea capensis, Paullinia fuscescens, P. nobilis, P. pinnata, P. trigonia, Pometia pinnata, Pseudima frutescens, Ratonia sp., Sapindus saponaria, Schleichera oleosa, Serjania arborea, S. brachystachya, S. caracassana, S. elegans, S. erecta, S. gracilis, S. paucidentata, S. plicata, S. schiedeana, S. trifoliata, S. trigretra, Talisia hemidasya, T. lexaphylla, Talisia sp.* aff. *pedicellaris, Talisia sp., Thouinia villosa, Thovinidium decandrum, Toechima dameliana, Tristiropsis acutangula, T. canarioides, Ungnadia speciosa*.

SAPOTACEAE
107 genera; 1,000 species

The members of this family are mostly tropical with a few temperate representatives. Several economic products come from the family: the latex is used in chewing gum; edible fruits, oils, guttapercha, and timber are also produced. A proteinaceous sweetener several times sweeter than sugar is also known from the family.

Several genera have been reported to contain alkaloids. Three species of these were encountered in this study: *Mimuspos elengi* (2/4), *Planchonella cotinifolia, P. thyrsoidea*. In addition, the following were positive: *Austromimusops* (= *Vitilliariopsis*) *dispor, Bequartiiodendron megalismontana, Bumelia lactovirens, B. verruculosa, Calocarpum sp., Chrysophyllum gonocarpum, Dipholis salicifolia* (2/3), *Chrysophyllum sp., Mastichodendron foetidissimum, Oxythece* (= *Neoxythece*) *elegans* (1/2), *Planchonella hochrenteineri, Planchonella spp.* (2/2), *Pouteria engleri, Sideroxylon capirii, S. guianensis.*

Negative tests were given by the following: *Achras sapota, Bassila latifolia, Bumelia celastrina, B. lanuginosa, B. occidentalis, Calocarpum mammosum, C. sapota, Chrysophyllum argenteum, C. balata, C. cainto, C. clausenii, C. fulum, C. mexicanum, C. pricurei, C. pulcherrimum, C. roxburghii, C. soboliferum, C. viridifolium, Ecclinusa sanguinolenta, Ecclinusa sp., Lucuma campechiana, L. palmeri, L. pariry, L. serpentaria, Lucuma sp., Madhuca indica, Manilkara bidentata, M. macaulayae, M. mochisia, M. paraensis, M. subserica, M. zapotilla, Martiusiella imperialis, Micropholis cruegeriana, Mimusops caffra, M. huberi, M. roxburghiana, M. rufula, M. siqueirae, M. zeyheri, Palaquium cf. lobbianum, Planchonella anteridifera, P. charactacea, P. firma, P. macropoda, P. myrsinoides, P. pohlmannianum, P. sarcospermoides, Plectocomiopsis geminiflorus, Pouteria caimito, P. elegans, P. guianensis, P. lasiocarpa, P. luzoniensis, P. maclayana, P. macrocarpa, P. macrophylla, P. megalismontana, P. multiflora, P. oppositifolia, P. pariri, P. reticulata, P. salicifolia, Pouteria spp.* (3), *P. speciosa, P. tora, P. virescens, Sarcaulus brasiliensis, Sideroxylon dulciferum, S. ferrugineum, S. inerme, S. meyeri, Sideroxylon sp., Sideroxylon tempisque.*

SARRACENIACEAE
3 genera; 15 species

These insectivorous plants are found in both eastern and western portions of North America as well as northeastern sections of South America.

In early reports, two species of *Sarracenia* gave positive tests for alkaloids but more recent references mention only the presence of amino acids.

Two samples representing two species of *Sarracenia*, *S. flava* and *S. leucophylla*, failed to give an indication of the presence of alkaloids.

SAURURACEAE
5 genera; 7 species

This small family is found in North America and eastern Asia. A substituted benzamide has been isolated from a species of *Houttuynia*, but in the present study no positive alkaloid tests were obtained on testing *Anemopsis californica*, *Houttuynia cordata*, *Saururus cernuus*, and *S. chinensis*.

SAXIFRAGACEAE
36 genera; 475 species

The taxonomy of this family appears to be quite complex. It has been maintained as a group; it has also been split into several families, some of which have, in turn, been further divided by other taxonomists. Those genera which have been assigned to the Grossulariaceae are treated separately in this account (see that family heading); those considered by some taxonomists to belong to the Hydrangeaceae are maintained here in the Saxifragaceae.

It is a subcosmopolitan family found mostly in north temperate and cold regions. A few cultivated ornamentals and edible fruits are its only economic importance.

Only three of 75 samples tested gave positive alkaloid tests: *Astilbe rivularis*, *Hydrangea arborescens*, *Vahlia capensis*.

The following were negative: *Astilbe chinensis*, *A. longicarpa*,

A. macroflora, Baurera rubioides, Bergenia crassifolia, B. ligulata,
C. major, Decumaria barbara, Deutzia corymbosa, D. crenata,
D. pulchra, D. scabra, D. taiwanensis, Heuchera americana,
H. mexicana, H. micrantha, Hydrangea angustipetala, H. arbores-
cens, H. aspera, H. bretschneideri, H. chinensis, H. hortensis,
H. integrifolia, H. longifolia, H. macrophylla, H. opuloides,
H. paniculata, H. petiolaris, H. quercifolia, H. radiata, H. serrata,
I. chinensis, I. oldhamii, I. virginica, Mitella breweri, M. formosa-
na, Montinia caryophyllacea, Penthorum sedoides, Philadelphus
coronarius, P. lewisii, P. madrensis, P. mexicanus, Phyllonoma lati-
cuspis, Polyosma rhytopholia, Saxifraga aizoon, S. decipiens,
S. hostii, S. pentadactylis, S. stolonifera, S. virginiensis, Schizoph-
ragma hydrangioides, S. integrifolia, Tiarella unifoliata, Vahlia di-
chotoma, Whipplea modesta.

SCROPHULARIACEAE
222 genera; 4,450 species

This is a cosmopolitan family noted for a number of ornamentals and for the famous cardiotonic digitalis.

Alkaloids have been reported in several genera and species, but they do not contribute to the chemistry of the family in a major way. Some of the plants parasitize those of other families and are known to take up the alkaloids of the latter (e.g., *Pedicularis*).

Positive tests were obtained in this study for the following: *Angelonia integerrima* (2/4), *A. salicariaefolia* (1/3), *Aptosimum calycinum, A. depressum, A. indivisum, A. leucorrhizum, A. lineare* (1/2), *A. organoides, A. spinescens, Bowkeria cymosa* (2/2), *Castilleja foliolosa, C. latifolia, C. mexicana, C. scorzoneraefolia* (1/4), *C. tenuifolia* (3/4), *Diascia capsularis, D. engleri, Diplacus (= Mimulus) calycinus, D. longiflorus, Elytraria acaulis, E. squamosa* (this genus name is included in either Acanthaceae or Scrophulariaceae by Willis, but in Acanthaceae only by Mabberley), *Freylinia lanceolata, Hebenstretia integrifolia* (in the Globulariaceae by Cronquist), *Lamourouxia multifida, L. pringleri* (2/2), *L. tenuifolia* (2/3), *Leucocarpus alatus, Leucophyllum virescens* (1/2), *Limnophila indica, Lindernia dubia, Macranthera flaminea, Manulea ob-*

ovata, M. parviflora, Maurandya (= *Asarina*) *antirrhinflora* (2/3), *M. erubescens, Moniera* (= *Bacopa*) *trifolia* (5/5), *Nemesia sp.*, *Pedicularis canadensis* (4/9), *P. mexicana, P. pectinata* (2/2), *Pentstemon antirrhinodes* (2/2), *P. apateticus* (1/3), *P. atrorubens, P. barbatus, P. campanulatum* (4/6), *P. centranthifolius, P. cordifolius, P. digitalis* (3/4), *P. eatonii, P. floridus, P. gentianoides* (1/3), *P. gracilentus, P. grinellii, P. heterophyllus, P. hidalgensis, P. hirsutus* (2/3), *P. kunthii* (4/4), *P. laetus* (2/3), *P. multiflorus, P. newberryi* (1/2), *Pentstemon sp., P. spectabilis, Polycarena cuneifolia, Russelia teres, Scoparia dulcis* (3/14), *Scrophularia californica, S. lanceolata* (1/2), *Seymeria pectinata, Stemodia viscosa, Striga asiatica* (1/2), *Sutera caerulea, S. glabrata, S. griquensis, S. integerrima, Verbascum olympicum, Wulfenia carenthica, Zaluzianskya maritima* (2/2). Alkaloids had previously been recorded in *Scoparia dulcis.*

Negative results were obtained with the following: *Adenosma glutinosa, Alectra kirkii, A. senegalensis, Anastrabe integerrima, Angelonia angustifolia, Angelonia sp., Antirrhinum coulteriana, A. majus, Aptosimum marlothii, Aureolaria flava, Bacopa caroliniana, B. elongata, B. monniera, B. procumbens, Buchnera cruciata, B. elongata, B. henriquesii, B. hispida, B. integrifolia, B. mexicana, B. tomentosa, B. virgata, Buttonia superba, Calceolaria sp., Capraria biflora, C. saxifrageofolia, Castilleja agrestis, C. angustifolia, C. applegatei, C. arvensis, C. breweri, C. disticha, C. douglasii, C. exilis, C. falcata, C. glandulosa, C. hirsuta, C. integra, C. integrifolia, C. latobracteata, C. laxa, C. lineariaefolia, C. lithospermoides, C. minata, C. plagiostoma, C. psittacina, C. schaffneri, Castilleja spp.* (2), *C. wightii, Chelone glabra, Collinsia bartsiaefolia, C. torreyi, Conobea scoparioides, Diascia integerrima, Digitalis purpurea, D. lanata, Erinus alpinus, Escobedia curialis, E. linearis, Esterhazya latifolia, E. splendens, E. splendida, Esterhazya sp., Euphrasia callosa, E. humifusa, Freylinia tropica, Gerardia communis, G. fascicularis, G. genistifolia, G. peduncularis, G. racemulosa, G. tenuifolia, Gerardia sp., Graderia scabra, Graderia sp., G. subintegra, Gratiola peruviana, G. virginiana, Halleria elliptica, H. lucida, Hebe odora, H. speciosa, H. stricta, H. tenuis, H. tetragona, H. venustula, Hebenstretia comosa, H. dentata, H. fruticosa, Hemiphragma heterophyllum, Herpestis monniera, Hyobanche sanguinea, Lamourouxia exserta, L. gracilis, L. lanceo-*

lata, L. rhinanthifolia, L. viscosa, Leucophyllum ambiguum, L. candidum, L. frutescens, L. griseum, L. minus, L. pringlei, L. revolutum, L. texanum, Limophila indica, L. rugosa, Linaria canadensis, L. cymbalaria, L. pinifolia, L. vulgaris, Lindenbergia polyantha, L. urticaefolia, Lindernia antipoda, Maurandya barclayana, M. erecta, Mazus japonicus, Mecardonia acuminata, Melampyrum lineare, Melasma hispidum, M. rhinanthoides, Microdon cylindricus, Mimulus aurianticus, M. bicolor, M. bigelovii, M. brevipes, M. cardinalis, M. glabratus, M. gracilis, M. moschatus, M. nepalensis, M. pilosus, M. puniceus, M. ringens, M. torreyi, M. viscidus, Mohavea confertiflora, Moniera cuneifolia, Morgania floribunda, Nelsonia canescens, Nemesia capensis, N. pubescens, Nemesia sp., Notochilus sp., Orthocarpus purpurascens, Parentucellia latifolia, Paulownia kawakamii, P. tomentosa, Pedicularis semibarbata, Pentstemon azureus, P. breviflorus, P. bridgesii, P. confusus, P. deustus, P. hallii, P. hartwegii, P. heterodoxus, P. incertus, P. isophyllus, P. pringlei, P. speciosus, P. tenuifolius, Physocalyx auranticus, P. major, Physocalyx sp., Russelia coccinea, R. cuneata, R. equisetiformis, R. floribunda, R. retrorsa, R. rotundifolia, R. sarmentosa, Russelia sp., R. syringaefolia, R. tenuis, R. tetrapetra, R. verticillata, R. villosa, Scoparia divaricata, Scrophularia californica, S. lanceolata, S. marilandica, Seymeria decurra, Seymeria sp., Sibthorpia pichinchensis, Sopubia delphinifolia, Stemodia hyptoides, S. macrantha, Striga bilabiata, S. gesnerioides, S. masuria, Sutera accrescens, S. albiflora, S. amplexicaulis, S. aspalathioides, S. atropurpurea, S. auriantica, S. burkeana, S. carvalhoi, S. crassicaulis, S. floribunda, S. grandiflora, S. hispida, S. micrantha, S. pinnatifida, S. revoluta, Sutera sp., Tetranema mexicana, Torrenia concolor, Tubiflora (= Elytraria) squamosa, Uroskinnera spectabilis, Verbascum blattaria, V. thapus, Veronica anagallis vaquatica, V. nivea, V. officinalis, V. peregrina, V. serpyllifolia, Veronicastrum virginicum, Zaluzianskya capensis.

SELAGINELLACEAE
1 genus; 600 species

The family is cosmopolitan but has no significant economic importance.

Alkaloids have not been discovered in the family. In the tests of the samples collected in this study, *Selaginella caudata, S. involvens,* and *S. longipinna* gave positive results while the remainder were negative: *S. arenicola, S. atroviridis, S. caudata, S. deliculata, S. deoderleinii, S. galeotii, S. involvens, S. lavordei, S. lepidophylla, S. leptophylla, S. longipinna, S. mollendorfii, Selaginella spp.* (2), *S. stenophylla, S. tamariscina.*

SIMAROUBACEAE
22 genera; 170 species

This is a tropical family with a couple of members in temperate Asia. Some are used medicinally, some for their oilseeds, some for timber, and some as ornamentals, especially *Ailanthus,* which has been introduced and naturalized in North America and central and southern Europe.

Alkaloids are known in the family including ∝-carbolines (harman) and canthinone. Of 26 species tested, the following were previously known to be alkaloidal and also gave positive tests in this study: *Ailanthus altissima* (2/4), *A. excelsa, Alvaradoa amorphoides* (1/3), *Picrasma javanica* (1/4).

The following also gave positive tests: *Quassisa andra, Simaba* (= *Quassia*) *cedron* (5/6), *Q. ferruginea, Simarouba* (= *Quassia*) *glauca* (1/2).

Negative results were obtained for *Alvaradoa humilis, Castela texana, C. tortuosa, Eurycoma apiculata, Kirkia acuminata, K. wilmsii, Picramnia macrostachya, P. pentandra, P. xalapensis, Picrodendron baccatum, Quassia indica, Recchia mexicana, Simaba* (= *Quassia*) *cuspidata, Simaba sp., Simarouba amara, Simarouba sp., S. versicolor,* and *Suriana maritima* (sometimes placed in Surianaceae).

SMILACACEAE
10 genera; 225 species

Tropical and warm zones, particularly of the southern hemisphere, are the major areas of distribution of this family, which has yielded sarsaparilla used as a medicinal tonic and as the base for refreshing drinks. Some species are also cultivated as ornamentals.

The family is not known for alkaloids. Two unidentified species of *Smilax* gave positive tests in this study, but 35 others did not, nor did *Behnia reticula, Geitonoplesium cymosum, Heterosmilax guadichaudiana,* and *Pleiosmilax vitensis.* The negative *Smilax* species included the following: *Smilax aequitorialis, S. bona-nox, S. bracteata, S. china, S. corbularia, S. cordifolia, S. cumanensis, S. elongatoreticulata, S. elongato-umbellata, S. glabra, S. glauca, S. herbacea, S. kraussiana, S. lancaefolia, S. laurifolia, S. medica, S. mexicana, S. moranensis, S. oldhami, S. opaca, S. oxyphylla, S. papyracea, S. racemosa, S. randaiensis, S. sandwicensis, Smilax spp.* (8), *S. spinosa, S. tortipetiolata.*

SOLANACEAE
90 genera; 2,600 species

This is a subcosmopolitan family especially rich in species in Andean South America. It furnishes a number of well-known drugs and tobacco, tomatoes, potatoes, peppers, hallucinogens used in aboriginal societies, as well as several common ornamentals.

Among the many chemical alkaloid types found in the family are alkaloids of tropane, steroid, $\propto$-carboline, nicotine, hygrine, and other structures responsible for many of the uses of the medicinal and hallucinogenic species of the group.

Many of the well-known alkaloidal representatives of the family were encountered in this study: *Atropa belladonna* (root) (2/2), *Capsicum frutescens* (2/3), *Datura arborea* (3/3) (the tree *Daturas* are now assigned to the genus *Brugmansia*), *D. candida* (3/3), *D. inoxia, D. leichardtii, D. metaloides* (4/4), *D. sanguinea* (4/4), *D. stramonium* (2/4), *D. suaveolens* (2/2), *Duboisia myoporoides, Fabiana imbricata* (2/2), *Lycium halmifolium, Nicandra physaloides, Nicotina glauca* (7/8), *N. repandra* (2/2), *N. tabacum, N. trigonophylla* (5/5), *Physalis pubescens, Solanum angustifolium, S. auriculatum, S. dulcamara* (2/2), *S. giganteum* (2/2), *S. gracile, S. incanum, S. jasminoides, S. macranthum* (2/2), *S. melongena, S. nigrum* (4/13), *S. nodiflorum* (1/1), *S. ovalifolium, S. paniculatum* (1/2), *S. rostratum* (2/4), *S. rugosum* (2/3), *S. seaforthianum* (2/2), *S. sodomeum* (1/2), *S. sturtianum, S. torvum* (2/4), *S. tuberosum, S. umbellatum* (2/2), *S. verbascifolium* (3/12), *S. xanthocarpum, Withania somnifera.*

Numerous other species were alkaloid-positive: *Brachistus diversifolius, B. pringlei, Brugmansia* (see *Datura*), *Brunfelsia americana, B. australis, B. densiflora, B. grandiflora, B. jamaicense, B. lactea, B. portoricensis, Brunfelsia spp.* (2), *Capsicum minimum, Cestrum benthamii* (1/4), *C. fasciculatum, C. laevigatum, C. orchiaceum, Cestrum sp., C. sendternianum* (1/2), *C. terminale* (1/4), *Chamaesaracha coronopus* (2/2), *Datura chlorantha, D. discolor* (3/3), *D. metel* (4/4), *Datura spp.* (3/4), *D. vulcanicola* (2/2), *Iochroma fuchsioides* (2/2) [this has been shown to be a false positive due to the structure(s) of the withanolides isolated from this plant (Raffauf, Shemluck, and LeQuesne, 1991)], *Iochroma sp., I. umbrosa, Juanulloa ochracea, Lycium acutifolium, L. austrinum, L. campanulatum, L. cooperi, L. hirsutum, L. kraussii, Lycopersicon esculentum, Margaranthus solanaceus* (1/2), *Melanantha sp., Methysticodendron amesianum* (2/2), *Nicotiana attenuata, Nicotiana sp., Petunia rupestris, Physalis acuminata* (1/4), *P. barbadensis, P. coztomatl, Physalis spp.* (3/4), *P. subglabrata, Solanum aculeastrum, S. antillarum, S. auriculatum* aff. *toxicarum, S. amazonicus, S. andrieuxii, S. apoporanum, S. bahamense, S. bifurcum, S. blodgettii, S. campaniforme, S. campylacanthum* (2/2), *S. cardiophyllum, S. caribbaeum, S. centrale, S. cervantesii* (4/4), *S. coccineum, S. crinitipes* (2/2), *S. diphyllum, S. dunalianum* (2/2), *S. erianthum* (2/2), *S. gemellum* (3/3), *S. hispidum, S. inaequale* (2/2), *S. jamaicensis, S. japonense, S. lacerde* (leaf, stem, fruit), *S. lycocarpum, S. lyratum* (1/2), *S. mammosum* (2/2), *S. marginatum, S. maximowiczii, S. megalochiton* (stem, bark, and wood), *S. medula, S. nyctaginoides, S. omitiomirense, S. photeinocarpum, S. pabstii, S. sanctae, Solanum spp.* (30/57), *S. subinerme* (stem), *S. swartzianum* (2/2), *S. woodburyii, S. xantii* (1/2), *Withania ashwaganda.*

Negative tests were obtained with the following species: *Bassovia lucida, Bassovia sp.* aff. *mexicana, Bouchetia erecta, Brunfelsia bonodora, B. guianensis, B. pauciflora, Brunfelsia spp.* (2), *Capsicum lucidum, C. mirabile, Capsicum sp., Cestrum amictum, C. anagyris, C. auriantiacum, C. diurnum, C. fasciculatum, C. flavescens, C. laevigatum, C. laxum, C. nocturnum, C. schicolei, C. strigillatum, Chamaesachara cornioides, Cyphomandra endopogon, C. sciadostylis, Cyphomandra sp., Datura alba, Datura ferox, Guilfoylia monostylis, Juanulloa mexicana, Lycium albiflorum, L. andersonii,*

L. barbinodum, L. berlandieri, L. carolinianum, L. chinense, L. ferocissimum, L. pallidum, Markea longipes, Markea sp., *Metternichia princeps, Nectouxia formosa, Nicotiana velutina, Nierembergia angustifolia, N. stricta, Petunia ericifolia, P. grandiflora, P. linearis, P. paraensis, P. regnellii, Petunia spp.* (2), *Physalis angulata, P. foetens, P. hederaefolia, P. minima, P. peruviana, Schwenkia divaricata, Schwenkia spp.* (4), *Sessea brasiliensis, Solandra guttata, S. nitida, S. aculeatissimum, Solanum biflorum, S. caroliniense, S. cernuum, S. chloropetalum, S. decorticans, S. diversifolium, S. douglasii, S. elaeagnifolium, S. ellipticum, S. grandiflorum, S. heterodoxum, S. hirtellum, S. indicum, S. kionotrichum, S. mitlense, S. ochraceoferrugineum, S. paraense, S. pensile, S. ratonetii, S. rigescens, S. stoloniferum, S. subinerme, S. tequilense, S. variable.*

REFERENCE

Raffauf, R. F., M. J. Shemluck, and P. W. LeQuesne, *Journal of Natural Products 34* (1991) pp. 1601-1606.

Many of the tests reported here were conducted not only to determine the presence of alkaloids (known to be present in many of the species of *Solanum*) but for those steroidal alkaloids potentially useful as starting materials for the production of steroid intermediates of medicinal importance. These tests were conducted by Dr. Melvin Shemluck at Northeastern University and Dr. Kazuko Kawanishi, Women's College of Pharmacy, Kobe, Japan. Their assistance in this portion of the screening is gratefully acknowledged as is the hospitality of the University of Puerto Rico, where some of the work was done, and the courtesy of INEXA (Industria Extractadora S.A.), Quito, Ecuador, which supplied financial support.

SONNERATIACEAE
2 genera; 7 species

A tropical Old World family, its habitats are the mangrove and rain forest areas. It is related to the Lythraceae. Alkaloids have been recorded for *Sonneratia* but they have not yet been characterized.

Two species of *Sonneratia* gave positive tests for alkaloids, *S. alba* (1/6) and *S. caseolaris* (3/6), while *Crypteronia parvifolia* (in a separate family, Crypteroniaceae, by some authorities), *Duabanga grandiflora, D. moluccana, D. sonneratioides, Sonneratia acida, S. apetala, S. griffithii, S. lanceolata, S. ovata,* and *S. pagtapata* did not.

SPARANGIACEAE
1 genus; 12 species

The family is found in the north temperate zone and south to Australia and New Zealand.

Sparangium has been reported to contain alkaloids in a couple of species; tests of two samples of *S. americanum* were negative.

SPHENOCLEACEAE
1 genus; 2 species

This small tropical family has been used as a potherb in Indonesia but was found to be toxic to cattle in areas of Amazonia where cows had taken it as fodder. No reports of such toxicity are known from areas of the Gulf Coast of the United States where it has been reported to be eaten by stock.

Little is known of the chemistry of the family (Raffauf and Higurashi, 1988). *Sphenoclea zeylanica* gave one positive test for alkaloids during the testing of five samples.

REFERENCE

Raffauf, R. F. and A. Higurashi, *Revista de la Academia Colombiana de Ciéncias Exactas, Fisicas y Naturales 16* (1988) pp. 99-105.

STACHYURACEAE
1 genus; 5-6 species

This Asian family extends from Japan to the Himalayas.

Alkaloids are not known. A sample of *Stachyurus himalaicus* gave a negative result when tested in this survey.

STACKHOUSIACEAE
3 genera; 28 species

This small family is found in Australasia: Malaysia, Australia, New Zealand, and the islands of the Pacific.

Alkaloids are not known in this family; three samples of *Stackhousia* (*S. intermedia* and *S. monogyna* [2]) were negative as tested in this study.

STAPHYLEACEAE
5 genera; 27 species

The presence of alkaloids has been reported in two genera of the family (*Staphylea, Turpinia*), but tests on *S. bumalda, S. pinnata, S. trifolia, Turpinia brachypetala, T. formosana, T. nephalensis, T. occidentalis,* and *T. pentandra* were without positive result.

STEMONACEAE
4 genera; 32 species

This family is found in eastern Asia, Indomalaysia, and south to tropical Australia as well as in eastern North America.

The genus *Stemona* is the source of insecticidal principles; alkaloids have been reported from *Stemona* and *Croomia*. Three species of *Stemona* (*S. australiana, S. sessilifolia,* and *S. tuberosa*) gave positive tests; the last two were known earlier.

STERCULIACEAE
72 genera; 1,500 species

Flourishing in warm and tropical areas with a few in temperate zones, the Sterculiaceae yield chocolate (cacao), cola nuts, timber, and cultivated ornamentals.

Many genera and species contain caffeine and other xanthines. Pyrrolidones, pyridones, and polypeptide alkaloids are also found in the family.

Positive tests were obtained for the following species: *Argyrodendron trifoliolatum, Ayenia fruticosa, Byttneria aculeata* (3/3), *B. herbacea, Cistanthera* (= *Neogordonia*) *papaverifera, Dombeya kirkii* (1/2), *Hermannia boraginifolia* (1/2), *H. brynifolia, H. burchellii, H. grandiflora, H. helianthemum* (2/2), *H. inflata, H. linearifolia* (2/2), *H. linearis, H. quartiniana, Hermannia sp.* (1/2), *H. spinosa, H. tomentosa* (1/2), *Mehlania rehmannii* (2/2), *Melochia corchorifolia* (previously known) (2/2), *M. nodiflora* (1/2), *M. pyramidata* (1/2) (previously known), *M. tomentosa* (1/3), *Pterospermum acerifolia, Sterculia alata* (previously known), *Waltheria americana* (1/3), *W. indica* (2/7), *Waltheria sp.* (1/2).

Negative tests were obtained with the following: *Abroma augusta, Argyrodendron actiniphylla, A. peralta, Ayenia sp., Brachychiton diversifolia, Byttneria scalpellata, Byttneria sp.* (4), *Chiranthondendron pentadactylon, C. platanoides, Cola acuminata, Commersonia partramia, C. echinata, Dombeya burgessiae, D. cayeuxii, D. claessensii, D. cymosa, D. rotundifolia, D. wallichii, Firmiana papuana, F. platinifolia, F. simplex, Fremontia* (= *Fremontodendron*) *californica, Guazuma ulmifolia, Helicteres angustifolia, H. guazumaefolia, H. lhotzkyana, H. mexicana, H. ovata, H. pentandra, Heritiera littoralis, H. paralta, Heritiera spp.* (2), *H. trifoliata, Hermannia alnifolia, H. brachymalla, H. candicans, H. candidissima, H. comosa, H. cuneifolia, H. depressa, H. desertorum, H. flammea, H. floribunda, H. geniculata, H. gerrardii, H. glanduligera, H. holosericea, H. hyssopifolia, H. lacera, H. mariae, H. modesta, H. mucronulata, H. pallens, H. prismatocarpa, H. resediflora, H. saccifera, H. salvifolia, H. ternifolia, H. tigreensis, H. transvaalensis, H. trifurca, H. vernicata, Kerandrenia integrifolia, Kleinhovia hospita, Melhania acuminata, M. didyma, M. forbesii, M. linearifolia, M. prostrata, M. randii, Melochia aculeata, M. concatonata, Melochia sp., M. umbellata, Pterocymbium beccarii, P. heterophyllum, Scaphium beccarianum, Sterculia caribbaea, S. clemensiae, S. colorata, S. cowentzii, S. fanaiho, S. laurifolia, S. murex, S. nobilis, S. quadrifida, S. quinqueloba, S. rogersii, Sterculia sp., S. urens, S. villosa, Theobroma bicolor, T. cacao, T. grandiflora, T. microcarpum, T. obovatum, Theobroma spp.* (2), *T. speciosa, Thomasia petalocalyx, Waltheria brevipes.*

STYLIDACEAE
5 genera; 150 species

This is a family of southern and southeastern Asia, Australasia, and South America. It furnishes a few ornamentals.

Alkaloids are not known in the family. *Stylidium graminifolium* and *S. laricifolium* were negative as tested in this study.

STYRACACEAE
12 genera; 165 species

The family is found in the warm temperate and tropical areas of the Americas, the Mediterranean, Southeast Asia, and western Malaysia.

The resins, especially of *Styrax*, have had a long history of medicinal use.

A positive alkaloid test had been reported in *Styrax*, but little more is known of the alkaloidal nature of the family.

In this study, positive tests were obtained on four species: *Halesia monticola, Styrax latifolium, S. serrulatum, S. suberifolium.*

Negative tests were obtained with the following: *Alniphyllum fortunei, A. pterospermum, Halesia carolina, H. diptera, Styrax americanum, S. camporum, S. formosanum, S. glabrescens, S. hookeri, S. japonicum, S. leprosus, S. martii, S. obassia, S. ramirezii, S. rigidifolium, S. serrulatum virgatum, Styrax spp.* (3), *S. suberifolium.*

SYMPLOCACEAE
1 genus; 250 species

This is a family of tropical and warm America and the eastern Old World except Africa. Occasional positive alkaloid tests had been reported earlier; two were obtained in this study: *Symplocos konishii, S. morrisonicola.*

The following were negative: *Symplocos* aff. *flavifolia, S. apolis, S. arisanensis, S. celastrinea, S. congesta, S. glauca, S. guianensis,*

S. lanceifolia, S. laurina, S. laxiflora, S. lucida, S. modesta, S. itatiaiae, S. paniculata, S. phyllocalyx, S. prionophylla, Symplocos spp. (6), *S. ternifolia, S. theaefolia, S. tinctoria, S. wikstroemiifolia.*

T

TACCACEAE
1 genus; 10 species

This small tropical family produces tubers which are used in some regions as food after the removal of bitter substances.

A few species of the genus have been reported to give positive tests for alkaloids, but three species tested in this study were negative: *Tacca chantrieri, T. leontopetaloides, T. pinnatifida.*

TAMARICACEAE
5 genera; 78 species

This Eurasian-African family extends chiefly from the Mediterranean to Central Asia. *Tamarix* has been called "manna" along with several other plants which yield mannitol and is not to be confused with biblical manna, possibly a lichen.

Positive alkaloid tests have been recorded for species of *Tamarix* and *Reaumuria*, but no specific compounds have been isolated or identified.

Seven samples of seven species of *Tamarix* were tested without positive result: *T. aphylla, T. austro-africana, T. chinensis, T. dioica, T. gallica, T. tetrandra, T. usneoides.*

TAXACEAE
6 genera; 20 species

This family of evergreens is found in mainly temperate regions, with one species as far south as New Caledonia. It is the source of the currently important taxol, a drug of potential use in the treatment of certain cancers. The trees otherwise serve as timber and ornamentals.

The genus *Taxus* has given positive alkaloid tests and taxine has been characterized.

In the study reported here, *Taxus baccata*, *T. canadensis* (2/2), *Taxus sp.*, and *T. speciosa*, all previously recorded as giving positive alkaloid tests, were positive. *Phyllocladus alpinus* (now assigned to a unigeneric family, Phyllocladaceae) and *Torreya nucifera* were also positive.

The following were negative: *Dacrydium araucaroides*, *D. bidwilli*, *D. cupressinum*, *D. elatum*, *D. laxifolium*, *Phyllocladus glaucus*, *P. trichomanoides*, *Taxus baccata pendula*, *Torreya taxifolia*.

TAXODIACEAE
10 genera; 14 species

The family appears to have an uneven distribution, being found in North America, central Asia, and Tasmania. Any economic importance is due to timber and cultivated ornamentals.

One genus, *Athrotaxis*, contains alkaloids of the homoerythrina type. Tests conducted here on 13 samples representing eight other species were negative: *Cryptomeria japonica*, *Metasequoia sp.*, *Sequoia sempervirens*, *Sequoiadendron gigantea*, *Taiwania cryptomerioides*, *Taxodium ascendens*, *T. distichum*, *T. mucronatum*.

Two of five samples of *Sciadopitys verticillata* were positive.

THEACEAE
28 genera; 520 species

This is essentially a tropical family with a few representatives found in warm temperate areas. It has been divided into several subfamilies which are treated as separate families by some taxonomists.

It is economically important as the source of tea (*Camellia sinensis*), some timber, and several ornamentals.

Alkaloids have been reported for the family and, of course, tea is known for its content of caffeine. In the present study, three species gave positive tests for alkaloids: *Eurya japonica* (1/6), *Gordonia axillaris, Ternstroemia cherryi* (sometimes in a family of its own, Ternstroemiaceae). The xanthines do not give good tests with the Dragendorff reagent.

It may be for this reason that the greater number of tests made in this study were negative: *Adinandra formosana, A. milletii, Adinandra spp.* (2), *Bonnetia sessilis, Camellia brevistyla, C. caudata, C. caudata gracilis, C. japonica, C. nokoensis, C. sasanqua, C. sinensis, Cleyera integrifolia, C. japonica, C. theaoides, Eurya acuminata, E. albiflora, E. brassii, E. chinensis, E. crenatofolia, E. glaberrima, E. leptophylla, E. rengechiensis, Eurya sp., Gordonia anomala, G. lasianthus, Laplacea* (= *Gordonia*) *fruticosa, Neottia sp., Pseudoeurya* (= *Eurya*) *crenatifolia, Sakakia* (= *Eurya*) *pseudocamellia, Schima sp., S. superba, Stewartia ovata, S. pseudocamellia, Taonabo* (= *Ternstroemia*) *occarpa, Ternstroemia brasiliensis, T. gymnanthera, T. japonica, T. pringlei, Ternstroemia sp., T. sphondylocarpa, T. sylvatica, T. tepezapote, Thea* (= *Camellia*) *sinensis, Tutcheria* (= *Pyrenaria*) *shinkoensis.*

THEOPHRASTACEAE
5 genera; 90 species

This is an American family supplying timber and some fish poisons.

A positive alkaloid test for a species of *Jacquinia* has been recorded, but four species of that genus and one of *Clavija* were negative in the tests conducted here: *Clavija lancifolia, Jacquinia armilaris, J. annularis, J. auriantica, J. revolata.*

THYMELAEACEAE
50 genera; 720 species

The family is cosmopolitan but especially prevalent in Australia and tropical Africa. Many species are toxic due to their content of coumarin glycosides. Economically important species contribute timber, incense, fiber, and a few ornamentals.

Some genera/species have been reported to give positive tests for alkaloids, but the family is not known for these constituents. In this survey of 84 samples including 74 species, the following gave positive tests: *Arthrosolen* (= *Gnidia*) *polycephalus*, *Daphne mezereum* (previously known), *D. retusa*, *Lachnaea densiflora*, *Lagetta lintearia*, *Lasiosiphon* (= *Gnidia*) *sp.*, *Passerina vulgaris*, *Pimelea collina*, *Struthiola dodecandra*, *S. leptantha*, *S. myrsinites*.

The following were negative: *Arthrosolen caleopehalus*, *A. gymnostachys*, *A. microcephalus*, *A. passerina obtusifolia*, *A. sericocephalus*, *Dais cotonifolia*, *Daphne arisanensis*, *Daphnopsis americana*, *D. beta*, *D. bonplandiana*, *D. fasciculata*, *D. mollis*, *D. racemosa*, *D. salicifolia*, *Dirca palustris*, *Draptes ericoides*, *Edgeworthia garneri*, *E. papyrifera*, *Edgeworthia sp.*, *Englerodaphne* (= *Gnidia*) *ovalifolia*, *Gnidia capitata*, *G. kraussiana*, *G. phyllodinea*, *G. chrysantha*, *G. chrysophylla*, *G. coriacea*, *G. geminiflora*, *G. oppositifolia*, *G. polystachya*, *G. sericeae*, *Gnidia sp.*, *Lasiosiphon burchellii*, *L. caffer*, *L. deserticola*, *L. kraussianus*, *L. meisnerianus*, *L. polyanthus*, *L. splendens*, *Leucosmis* (= *Phaleria*) *sp.*, *Lophostoma callophylloides*, *Passerina montana*, *P. obtusifolia*, *P. paleacea*, *P. rigida*, *P. rubra*, *Peddiea africana*, *Phaleria octandra*, *P. sorgerensis*, *Pimelea altior*, *P. congesta*, *P. decora*, *P. lingustrina*, *P. octophylla*, *P. serpyllifolia*, *Schoenobiblus grandifolia*, *Struthiola argentea*, *S. ciliata*, *S. hirsuta*, *S. parviflora*, *Wikstroemia australis*, *W. indica*, *W. mononectaria*, *W. oahuensis*.

TILIACEAE
48 genera; 725 species

This subcosmopolitan family produces one economically important fiber, jute, as well as some timber and local medicinals.

Alkaloids are known in the family, including harman derivatives in *Grewia*. In this survey, the following were positive: *Grewia microcos* (2/2), *G. vernicosa*, *Microcos paniculata* (1/2), *Triumfetta bartramia* (1/2).

The remaining species tested were negative: *Althoffia* (= *Trichospermum*) *pleiostigma*, *Apeiba aspera*, *A. schomburgkii*, *Belotia* (= *Trichospermum*) *mexicana*, *Carpodiptera* (= *Berrya*) *ameliae*, *Corchorus acutangulus*, *C. capsularis*, *C. confusus*, *C. kirkii*,

C. olitorius, C. pasacuorum, C. saxatilis, C. siliquosus, C. tridens, C. trilocularis, C. velutinus, Grewia avellana, G. bicolor, G. biloba, G. glabrescens, G. caffra, G. decemovulata, G. falcistipula, G. flava, G. flavescens, G. lasiocarpa, G. latifolia, G. laevigata, G. lepidopetala, G. monticola, G. occidentalis, G. pachycalyx, G. rogersii, Grewia robusta, Grewia spp. (2), *G. stolzii, G. subpathulata, G. sulcata, G. vernicosa, G. villosa, Heliocarpus donell-smithii, H. occidentalis, H. pallidus, H. popayanensis, H. terebinthaceus, H. velutinus, Huntingia calabura, Hydrogaster sp., H. trinervia, Luehea candida, L. divaricata, L. speciosa, Luhea sp., L. uniflora, Mollia speciosa, Sparmannia ricinocarpa, Tilia americana, T. cordata, T. heterophylla, T. houghi, T. japonica, T. miqueliana, T. occidentalis, Tilia sp., Triumfetta althaeoides, T. angolensis, T. annua, T. brevipes, T. dekindtiana, T. digitata, T. discolor, T. dumetorum, T. falcifera, T. glechomoides, T. grandiflora, T. heliocarpoides, T. obovata, T. pilosa, T. polyandra, T. rhomboidea, T. semitriloba, T. sonderi, Triumfetta spp.* (3), *T. tenuipedunculata, T. tomentosa, T. welwitschii, T. winneckiana.*

TREMANDRACEAE
3 genera; 43 species

This small family of temperate Australia is chemically unknown. A sample of *Tetratheca thymifolia* gave a negative test for the presence of alkaloids.

TRIGONIACEAE
3 genera; 26 species

The Trigoniaceae constitute a tropical family of rather uneven distribution having one genus in Madagascar, one in tropical America and one in western Malaysia.

Alkaloids are known, but a sample of *Trigonia pubescens* and two of *T. nivea* were negative.

TROCHODENDRACEAE
1 genus; 1-2 species

Euptelea, placed by some taxonomists in its own family (Eupte-
leaceae) and *Trochodendron* are found in Asia from Korea and
Japan to Taiwan. Isothiocyanates are present, but alkaloids have not
been reported. A sample of *E. polyandra* gave a positive test; two
samples of *Trochodendron aralioides* did not.

TROPAEOLACEAE
3 genera; 88 species

These are herbs of Central and South America and alkaloid-negative as far as is known. However, one sample of *Tropaeoleum majus*
gave a positive test.

TURNERACEAE
10 genera; 110 species

This family of the warm and tropical areas of America and Africa
contains a few alkaloid-positive species, including *Turnera*, which
contains caffeine.
Turnera ulmifolia, known to be alkaloidal, was also found to be
so in this study (1/4), while other representatives of the family were
negative: *Piriqueta capensis, P. caroliniana, P. cistoides, Piriqueta
sp., Turnera diffusa, Turnera heterophylla, Turnera sp., Wormskiol-
dia lobata, W. longipedunculata.*

TYPHACEAE
1 genus; 10-12 species

These are cosmopolitan marsh plants which have been used in
local construction as well as food and medicine.
Positive tests for alkaloids have been noted for a few species, but
apparently they have not yet been characterized. Ten samples of

four species of *Typha* were negative in the tests conducted here: *T. angustifolia, T. capensis, T. dominguensis, T. latifolia.*

U

ULMACEAE
16 genera; 140 species

Trees of this family occur from tropical to chiefly north temperate areas. They are noted for timber (elm), a medicine (slippery elm bark), and ornamentals.

Occasional reports of the presence of alkaloids have appeared and alkaloidlike substances have been noted in a few of the species. In this study, 56 samples representing 33 species were tested and gave but three positive results: *Celtis formosana* (1/2), *C. mississippiensis, Ulmus americana* (1/2). The rest of the samples were negative: *Ampelocera hottlei, Celtis africana, C. ambylophylla, C. caudata, C. iguanea, C. laevigata, C. monoica, C. occidentalis, C. pallida, C. reticulata, C. sinensis, C. triflora, Chaetachme aristata, Gironniera celtidifolia, Hemiptelea davidii, Myriocarpa longipes, Trema amboinensis, T. cannabina, T. floridana, T. guineensis, T. micrantha, T. orientalis, Ulmus fulva, U. parvifolia, U. serotina, Ulmus sp., U. thomasii, U. uyematsui, Zelkova serrata.*

UMBELLIFERAE
418 genera; 3,100 species

The family is cosmopolitan with concentration in areas of the north temperate zone and tropical mountains. They are mostly herbs important as foods and flavors (celery, carrots, fennel, caraway, aniseed, etc.); *Conium* is toxic.

A few members contain alkaloids and several of these have been characterized. Only a few of the species which gave positive tests in this study were known to be alkaloidal from previous studies: *Ammi majus, Apium leptophyllum* (1/3), *Cicuta douglasii, Conium maculatum, Sanicula marilandica* (2/2).

Other positive species included: *Angelica venuosa, Anethum graveolens, Archangelica* (= *Angelica*) *gmelini, Cicuta maculata* (1/2), *Eryngium bourgati, Heteromorpha arborescens, H. trifoliata* (1/3), *Lefebvrea sp., L. welwitchii, Lichtensteinia lacera, Peucedanum capense, P. galbanum, P. multiradiatum, Rhyticarpus difformis, Sanicula canadensis, Selinum tenuifolium, Steganotaenia arilacea* (1/2), *Taushia hartwegii, Thaspium barbinode* (9/11).

The following were negative: *Alepidea amatymbica, A. gracilis, A. setifera, Apium graveolens, A. prostratum, Apium sp., Arctopus echinatus, Arracia atropurpurea, Arthriscus scandicina, Berula thunbergii, Bupleurum kaoi, B. munotii, Caucalis pedunculata, Centella asiatica, C. coriacea, C. flexuosa, C. fusca, C. glabrata, C. madagascariensis, C. virgata, Chaerophyllum tainturieri, Chammaele decumbens, Cicuta mexicana, Cnidium formosanum, Conium chaerophylloides, Cryptotaenia canadensis, Daucus carota, D. pusillus, Diplolophium swynnertonii, D. zambesianum, Eryngium aquaticum, E. aromaticum, E. beecherianum, E. bonplandii, E. canaliculatum, E. carlinae, E. chamissonis, E. foetidum, E. frotoeflorum, E. ghiesbregtii, E. gracile, E. heterophyllum, E. monocephalum, E. pectinatum, E. protaeflorum, E. rojasii, E. serratum, Eryngium spp.* (13), *E. sparganophyllum, E. stenophyllum, E. yuccifolium, Foeniculum vulgare, Heracleum ianatum, Hermas depauperata, Heteromorpha involucrata, H. kassneri, H. transvaalensis, Hydrocotyle americana, H. asiatica, H. benguetensis, H. bonariensis, H. hirta, H. javanica, H. leucocephalia, H. ranunculoides, H. sibthorpoides, Hydrocotyle sp., Lefebvrea stuhlmannii, Lomatium mohavense, L. torreyi, Myrrhis odorata, Neonelsonia acuminata, Oenanthe benghalensis, O. linearis, Oreomyrrhis andicola, O. papuana, O. pumila, Osmorrhiza aristata, O. asiatica, O. claytoni, O. occidentalis, Oxypolis filiformis, O. greenmanii, Pastinaca sativa, Perideridia bolanderi, P. oregana, Petroselinum crispum, Peucedanum caffrum, P. eyelesii, P. formosanum, P. japonicum, Pimpinella bechananii, P. huillensis, P. nitakayamensis, Pituranthos aphyllus, Platysace valida, Pteryxia petraea, P. terebinthina, Ptilinium copillaceum, Sanicula elata, S. lamelligera, Sanicula sp., S. tuberosa, Selinum capitellatum, S. japonicum, Sium repandrum,*

S. suave, Taushia arguta, T. nudicaulis, T. parishii, Torilis japonicus, Trachymene adenodes, T. arfakensis, T. glaucifolia, T. novoguineensis, T. rosulans, T. saniculaefolia, Zizia aptera.

URTICACEAE
52 genera; 1,050 species

The family is essentially tropical with a few temperate species familiar as "stinging nettles," which often produce painful sensations on contact with the leaves. Otherwise fibers (ramie) and a few edible leaves are known in the family.

Not much is known of its chemistry. Positive tests for alkaloids have been obtained in several species: 5-hydroxytryptamine has been identified in *Urtica*; the substance responsible for the burning sensation from the leaves of *Laportea* is a complex octapeptide; piperidine derivatives are found in the family as well.

In the tests conducted on 102 samples of 74 species, the following were positive: *Boehmeria caudata* (2/3), *B. cylindrica* (previously known) (3/3), *Boehmeria sp.*, *B. spicata*, *B. ulmifolia*, *Coussapoa orthoneura*, *Coussapoa sp.*, *Cypholophus friesianus*, *Cypholophus sp.*, *Myrianthus arboreus*, *Parietaria pennsylvanica*.

Negative results were obtained on testing the following: *Boehmeria densiflora, B. nivea, B. platyphylla, B. ramiflora, B. rugulosa, B. spicata, Brosimum sp., Cecropia bureaniana, C. denepus, C. mexicana, C. peltata, C. schiediana, Cecropia spp.* (3), *Coussapoa asperifolia, C. intermedia, C. latifolia, C. schottii, Coussapoa sp., Cypholophus peltata, Debregeasia edulis, Elatostema herbaceifolium, Fleurya alatipes, F. grossa, F. interrupta, F. mitis, Forskahlea candida, Laportea crenulata, L. decumana, L. photinophylla, L. pustulosa, Leucosyke capitellata, Maoutia setosa, Missiessya sp., Myriocarpa longipes, Nanocnide japonicum, Naucheopsis* aff. *calsneura, Parietaria debilis, Pellea andromedaefolia, Pellonia radicans, P. scabra, Phenax sonneratii, Pilea brevicornuta, P. peploides, P. pubescens, P. pumila, Pilea spp.* (2), *P. stipulosa, P. trinervia, Pipturus albidus, Poikilospermum sp., Polionia scabra, Pourouma acutiflora, Pourouma cecropiaefolia, Pourouma spp.* (2), *Pouzolzia elegans, P. hypoleuca, P. palmeri, P. parasitica, P. pentandra, P. zeylanica,*

Procris laevigata, Rousselia humilis, Touchardia latifolia, Urera alceifolia, U. baccifera, U. caracasana, U. tenax, U. woodii, Urtica chamaedoides, U. dioica, U. pilulifera, U. spathulata, U. urens.

Cecropia, Coussapoa, Myrianthus, Poikilospermum, and *Pourouma* have been placed in a separate family, Cecropiaceae, by some authorities.

V

VALERIANACEAE
17 genera; 400 species

This is a cosmopolitan family but it is found especially in north temperate areas and the Andes of South America. Most are garden plants though some are edible, and *Valeriana* is medicinally used.

Iridoid alkaloids have been recognized in the family. In this study, the following were positive: *Nardostachys jatamansi* (2/2), *Valeriana capitata, V. ceratophylla, V. clematitis* (2/2), *Valeriana sp.* indet. (1/3), *V. urticaefolia* (1/2), *V. vaginata.*

Negative tests were obtained from the following: *Astrephia chaerophylloides, Centranthus ruber, Patrinia villosa, Triplostegia glandulifera, Valeriana mexicana, V. palmeri, V. scandens, V. sorbifolia, Valerianella sp.*

VELLOZIACEAE
6 genera; 252 species

Members of this family are found in South America, Africa, Madagascar, and southern Arabia.

Alkaloids are not known in the family. Positive tests were obtained here with *Vellozia humilis* and *V. lithophylla,* but the following species of the same genus were negative: *V. equisetoides, V. intermedia, V. retinervis, Vellozia spp.* (6).

VERBENACEAE
91 genera; 1,900 species

There are a few temperate representatives of this family, which is otherwise tropical. It is best known, perhaps, for teak and weedy plants (*Lantana*).

A variety of alkaloidal structures have been identified in the family, known positive plants having included *Citharexylum inerme* (1/4), *Duranta repens* (2/3), *Lantana camara* (2/14), *Verbena officinalis* (3/5), and *V. trifolia* (2/3), which were also positive when tested in this study. Other alkaloid-positive species were the following: *Aegiphila brachiata, Aloysia lycroides* (1/6), *Avicennia germinans, A. marina, A. nitida* (1/6), *Callicarpa japonica* (1/3), *C. pedunculata, Callicarpa sp.* (1/2), *Chascanum garipense, Chascanum sp., C. schlecteri, Citharexylum myrianthum* (1/3), *C. coccineum, C. glabrum* (1/2), *C. indicus* (1/2), *C. lanceolatum, C. trichotomum, Lantana aristata, Lippia callicorpaefolia* (1/3), *Lippia sp.* (1/12), *Newcastelia spodiotricha, Premna spinosa, Privia sp.* indet., *Stachytarpheta sp.* (1/2), *Verbena recta* (1/3), *Vitex angus-castus, V. amboinensis, V. lucens, V. welwitchii.*

A large number of the samples tested were negative: *Aegiphila filipes, A. integrifolia, A. intermedia, A. lhotskiana, A. obducta, A. perplexa, A. racemosa, Aegiphila spp.* (2), *A. verticillata, Aloysia oblongeolata, A. polygalaefolia, A. pulchra, A. virgata, Amasonia hirta, A. campestris, Avicennia officinalis, A. resinifera, A. schaueriana, Bouchea fluminensis, B. prismatica, Callicarpa americana, C. lourieri, C. nudiflora, C. pilosissima, C. randaiense, Caryopteris wallichiana, Chascanum hederaceum, C. latifolium, C. pinnatifidum, Chloanthes parviflora, Citharexylum affine, C. ellipticum, C. fruticosum, C. lucidum, C. lycoides, C. oleinum, C. ovatifolium, C. poeppigii, C. solanaceum, C. spinosum, Clerodendron buchneri, C. bungei, C. cryptophyllum, C. discolor, C. floribundum, C. fortunatum, C. fragrans, C. inerme, C. infortunatum, C. nutans, C. paniculatum, C. thomsonae, C. triphyllum, C. speciosissium, C. syphonanthus, C. tanganikense, C. toxicarium, C. ugandensis, C. wildii, Congea tomentosa, C. ventulina, Cornutia grandifolia, Dicrastylis exsuccosa, Duranta sp., D. stenophylla, D. vestita, Faradaya splendida, Gmelina arborea, G. elliptica, G. moluccana, G. smithii,*

Holmskioldia sanguinea, H. spinescens, H. tettensis, Lantana achyranthifolia, L. fucata, L. hispida, L. horrida, L. involucrata, L. lockhartii, L. macropoda, L. montenidensis, L. rugosa, L. salvifolia, L. sellowiana, Lantana spp. (11), *L. tiliaefolia, L. trifolia, L. velutina, Lippia arborea, L. barbata, L. berlandieri, L. chrysantha, L. citriodora, L. grata, L. graveolens, L. hirta, L. hypoleia, L. javanica, L. lupulacea, L. lupulina, L. micromera, L. nodiflora, L. obscura, L. rehmannii, L. scaberrima, L. turnerifolia, L. umbellata, L. wilmsii, Oxera sp., Petrea arborea, Petrea zanguebarica, P. volubilis, Phyla betulifolia, Premna barbata, P. herbacea, P. integrifolia, P. modiensis, Priva grandiflora, P. lappulacea, P. mexicana, Pygmaeopremna (= Premna) sessilifolia, Stachytarpheta acuminata, S. albiflora, S. cayennensis, S. coccinea, S. jamaicensis, S. glabra, S. indica, S. jamaicensis, S. lactea, S. maxmilliani, S. mutabilis, S. sellowiana, S. trispicata, S. urticaefolia, Stilbe ericoides, Tectona grandis, Teijsmanniodendron hollrungii, Verbena ambrosifolia, V. amoena, V. bonariensis, V. brasiliensis, V. caroliniana, V. ciliata, V. dusenii, V. elegans, V. halei, V. hastata, V. hirta, V. lasiostachys, V. littoralis, V. malmei, V. neomexicana hirtella, V. phloziflora, V. pumila, V. rigida, V. scabra, Verbena spp.* (7), *V. urticifolia, Vitex acuminata, V. altissima, V. capitata, V. cofassus, V. flavens, V. harveyana, V. helmsleyi, V. megapotamia, V. mollis, V. mombassae, V. montevidensis, V. negundo, V. orinocensis, V. parviflora, V. payos, V. polygama, V. pyramidata, V. quinata, Vitex spp.* (4), *V. spongiocarpa, V. wilmsii, V. zeyheri.*

VIOLACEAE
28 genera; 830 species

The violets are cosmopolitan with *Viola* temperate. They include a few sources of medicinals and perfume oils, but most are valued as garden plants.

Little chemical work has been done on the family. Reports of the presence of the alkaloids exist, and an odd amide has been identified in *Hybanthus*.

Tests were done on 67 samples representing 51 species with the following positive: *Alsodeia* (= *Rinorea*) *sp., Hybanthus calceolaria, H. polygalifolium, Viola glabella, Viola sp.* (1/4).

Negative tests were obtained with the following: *Amphirrox lon-gifolia, Anchietea salutaris, Corynostylis arborea, C. excelsa, C. hibanthus, Hybanthus bigibbosus, H. communis, H. enneasper-mus, H. filiformis, H. ipecacuanha, H. mexicanus, H. verbenaceus, H. verticillatus, H. yucatanensis, Hymenanthera latifolia, Melicytus lancefolatus, M. ramiflorus, Noisettia orchidiflora, Payparola grandiflora, P. guianensis, R. macrocarpa, Rinorea* aff. *macrocar-pa, R. guianensis, R. passoura, R. pilosula, Rinorea sp., Viola abys-sinica, V. arcuata, V. arvensis, V. betonicifolia, V. cerasifolia, V. diffusa, V. flabelliformis, V. grahami, V. inconspicua, V. kitaibe-liana, V. lagaipensis, V. lanceolata, V. oblonga, V. pedata, V. penn-sylvanica, V. primulifolia, V. saggitata, V. sororia, V. striata, V. taiwaniana.*

VITACEAE
13 genera; 800 species

This is essentially a family of tropical to warm regions with extension into temperate areas (grapes). The fermented juice of the grape has been known since ancient times and cultivated varieties of the fruit are many. Otherwise, the family is known for some house plants.

Positive tests for the presence of alkaloids have been recorded, but these are not significant contributions to the chemistry of the family.

In the studies reported here, the following gave positive tests: *Cissus sp.* (1/8), *Cyphostemma congestum, Leea manillensis* (now in a family of its own, Leeaceae).

Negative tests were given by the following: *Ampelocissus africa-na, A. obtusata, Ampelopsis arborea, A. brevipedunculata, A. can-toniensis, A. cordata, Cayratea gracilis, Cissus cornifolia, C. erosa, C. gongyloides, C. grisea, C. quadangularis, C. guerkeana, C. hae-matantha, C. humilis, C. intergrifolia, C. petiolata, C. rhodesiae, C. rhombifolia, C. rotundifolius, C. schmitzii, C. sicyoides, C. sim-siana, C. welwitschii, Cyphostemma anatomica, C. bororense, C. cirrhosum, C. crotalarioides, C. gigantophyllum, C. graniticum, C. kerkvoordei, C. lanigerus, C. milbraedii, C. obovato-oblongum, C. sandersonii, Cyphostemma sp., C. spinopilosum, C. subciliatum,*

C. vanneelii, C. viscosum, C. woodii, Leea coccinea, L. hirta-acuta, L. macrophylla, L. indica, L. sambucina, Parathenocissus quinquefolia, P. tricuspidata, Rhoicissus cuneifolia, R. digitata, R. revoilii, R. rhomboidea, R. tomentosa, R. tridentata, Tetrastigma gormosana, T. umbellata, T. vomerensis, Vitis auriculata, V. blanco, V. californica, V. flexuosa, V. latifolia, V. limnacea, V. quadrangularis, V. repens, Vitis spp. (2), *V. vulpina.*

VOCHYSIACEAE
7 genera; 210 species

This is a family of tropical America and West Africa. It is not of economic importance. Vochysine, a pyrrolidine-substituted flavonoid, has been identified in the fruits of five species of *Vochysia*.

Vochysia guianensis was found positive in this study while *Callisthene sp., Qualea acuminata, Q. albiflora, Q. cerulea, Qualea spp.* (3), *Salvertia comallariodora,* and the following species of *Vochysia* were not: *V. bifalcata, V. divergens, V. elliptica, V. inundata, V. polyantha, V. magnifica, V. rufa, Vochysia spp.* (18), *V. tucanorum, V. thyrsoidea.*

W

WINTERACEAE
5 genera; 60 species

The members of this family are scattered in South America, Australia, New Guinea, and the southwestern Pacific with one genus/species in Madagascar. One genus, *Drimys*, is cultivated, but rarely.

Two species of *Drimys* have been reported to contain alkaloids but four others (*D. buxifolia, Drimys spp.* [2], and *D. winteri*) along with *Bubbia sp.* and *Pseudowintera colourata* did not.

A sample of *Bubbia* (= *Zygogynum*) *argentea* gave a positive test.

X

XANTHOPHYLLACEAE
1 genus; 93 species

This monogeneric family is Indo-Malaysian. Alkaloids are not known nor were they found in three samples of *Xanthophyllum papuanum*.

XANTHORRHOEACEAE
9 genera; 60 species

The family is found from Australia to New Guinea and New Caledonia. Alkaloids are not known. *Lomandra banksii* and *L. diffusa* were negative.

XYRIDACEAE
5 genera; 260 species

There are a few temperate species in this family but it is otherwise confined to tropical and warm areas. Two species of *Xyris* are used as aquarium plants.

The chemistry of this small family is not well known; alkaloids have not been reported nor were they found in *Abolboda macrostachys, Abolboda sp., Xyris caroliniana, X. fimbriata,* and two unidentified species of *Xyris*.

Z

ZAMIACEAE
8 genera; 100 species

These cycads of the tropical and warm areas of America, Africa, and Australia contain a toxic principal macrozamine, which is carcinogenic.

Alkaloids were not detected in the following: *Bowenia spectabilis, Encephalartos altensteinii, E. laurifolius, Lepidozamia peroffskyana, Macrozemia hopei, Stangeria eriopus* (sometimes in a family of its own, Stangeriaceae), *Zamia sp., Zamia spartea.*

ZINGIBERACEAE
53 genera; 1,300 species

The ginger family is tropical and especially prominent in Indo-Malaysia. It is known for the familiar spice as well as aromatic oils. Some members are used as ornamentals.

Alkaloids have not been observed in the family. Positive tests given by the following are likely due, at least in part, to nonnitrogenous substances reacting with the Dragendorff reagent, which has been observed in many other cases: *Afromomum melegueta, Curcuma zoedaria* (1/2), *Hedychium coronarium* (1/4), *Zingiber cassumunar, Z. officinale.*

Negative tests were obtained with the following species: *Alpinia kelugensis, A. purpurata, Alpinia sp., A. speciosa, Costus brasiliensis, C. lasius, C. speciosus, Costus spp.* (3), *C. spicatus, Curcuma longa, Elatteria cardamomum, Globba marantina, Hornstedtia schottiana, Kaempferia rosea, Renealmia aromatica, R. exaltata, R. lativaginata, R. longipes, Zingiber sp.*

ZYGOPHYLLACEAE
27 genera; 250 species

Tropical and warm, especially arid regions are home to the Zygophyllaceae. They are sometimes halophilic. Timber (lignum-vitae), wax, some medicinals, dyes, and ornamentals constitute their economic importance.

Alkaloid tests previously reported positive for *Peganum mexicanum* (2/2) and *Porlieria angustifolia* (1/2) were confirmed in the tests reported here. In addition, the following positives were observed: *Balanites australis, B. maughamii, B. roxburghii, Guiacum sanctum, Larrea tridentata* (1/3), *Morkillia mexicana* (1/2), *Sisyn-*

dite spartea (1/2), *Tribulus excrucians, Tribulus sp., Tribulus terrestris, Zygophyllum gilfillani, Zygophyllum sp.*

Negative tests were given by the following: *Balanites aegyptica, Fagonia californica, Guiacum coulteri, Kallstroemia maxima, Larrea divaricata, Nitraria schoberi, Sericodes greggii, Tribulus cistoides, T. terrestris, T. zeyheri, Zygophyllum cuneifolium, Z. flexuosum, Z. fruticosulum, Z. fulvum, Z. iodocarpum, Z. morgsana, Z. simplex, Z. spinosum, Z. suffruticosum.*

Appendix

Samples tested in this study were deposited in or, in the case of common species, identified by the following herbaria:

AK   Auckland Institute and Museum, Auckland, New Zealand; South Pacific including Fiji, Samoa, New Caledonia, Lord Howe and Norfolk Islands.

AMAZ   Herbario, Universidad Nacional de la Amazonia Peruana, Iquitos.

AMES   Orchid Herbarium of Oakes Ames, Harvard University, Cambridge, Massachusetts.

ARN   Herbarium, Arnold Arboretum, Harvard University, Cambridge, Massachusetts.

BHCB   Herbário, Dept. de Botanica, ICB, Belo Horizonte, Brazil.

BISH   Herbarium, Bishop Museum, Honolulu, Hawaii.

CANB   Australian National Herbarium, Canberra.

COL   Herbário Nacional Colombiano, Instituto de Ciéncias Naturales, Bogota.

ENCB   Herbário, Instituto Politécnico Nacional de Mexico, Mexico City, D.F.

ENT   Herbarium, Forest Department, Entebbe, Uganda.

FLAS   Herbarium, University of Florida, Gainesville, Florida.

FTG   Herbarium, Fairchild Tropical Garden, Miami, Florida.

GH	Gray Herbarium of Harvard University, Cambridge, Massachusetts.
HK	Hong Kong Herbarium, Kowloon, Hong Kong.
IAN	Herbário, Instituto Nacional Agronômico do Norte, Belém, Brazil.
INPA	Herbário, Instituto Nacional de Pesquisas de Amazônica, Manaus, Brazil.
LA	Herbarium, Biology Department, University of California at Los Angeles, California.
MEXU	Herbário, Instituto Botánico, Universidad Nacional de Mexico, Mexico City, D.F.
MOAR	Herbarium, Morris Arboretum, Philadelphia, Pennsylvania.
NEBC	Herbarium, New England Botanical Club, Cambridge, Massachusetts.
NY	New York Botanical Garden Herbarium, Bronx, New York.
PRE	National Herbarium, Pretoria, South Africa.
RB	Herbário, Jardim Botanico, Rio de Janeiro, Brazil.
SING	Herbarium, Botanical Gardens, Singapore.
SJ	Herbário, Departmento de Recursos Naturales, Puerta de Tierra, Puerto Rico.
SRGH	National Herbarium and Botanical Garden, Harare, Zimbabwe.
TAN	Herbier, Department Botanique Antananarivo, Madagascar.
TRIN	National Herbarium, Trinidad and Tobago, St. Augustine, Trinidad.
UCWI	Herbarium, University of the West Indies, Mona, Kingston, Jamaica.

Collections made in southern Brazil are in Herbário Hatschbach, Curitiba, Parana, Brazil, a private herbarium not listed in Index Herbariorum. Some West African collections were made under the auspices of the Chelsea College of Science and Technology, London. Some of the southern Mexican plants were collected and identified by the late Donald Cox, Oaxaca.

Bibliography

Major references to plant sources of the known alkaloids.

Abisch, E. and T. Reichstein, *Helvetica Chimica Acta. 43* (1960) p. 1844.

Balick, M. J., L. Rivier, and T. Plowman, *Journal of Ethnopharmacology 6* (1982) p. 287.

Collins, D. J., C. C. J. Culvenor, J. A. Lamberton, J. W. Loder, and J. R. Price, *Plants for Medicines: A Chemical and Pharmacological Survey of the Plants of the Australian Region.* C.S.I.R.O. Australia, Melbourne, 1990.

Farnsworth, N. R., *Journal of Pharmaceutical Sciences 33* (1966) p. 225.

Gibbs, R. D., *Chemotaxonomy of Flowering Plants*, Vols. I-IV. McGill-Queen's University Press, Montreal, London 1974.

Habib, A. M., *Journal of Pharmaceutical Sciences 69* (1980) p. 37.

Hegnauer, R., *Chemotaxonomie der Pflanzen*, Vols. 1-9. Birkhauser Verlag, Basel, 1962-1990.

Holmgren, P. K, N. H. Holmgren, and L. C. Barnett, *Index Herbariorum*, 8th Edition, Part I. New York Botanical Garden, Bronx, NY 1990.

Lawrence, G. H. M., *Taxonomy of Vascular Plants*, Macmillan, NY, 1951.

Mabberley, D. J., *The Plant Book*, Cambridge University Press, 1989.

Philippson, J. D., *Phytochemistry 21* (1982) p. 2441.

Raffauf, R.F., *Economic Botany 16* (1962) p. 171.

_____, *Lloydia 25* (1962a) p. 255.

_____, *Handbook of Alkaloids and Alkaloid-Containing Plants.* Wiley Interscience, New York, 1970.

Raffauf, R. F., and S. von R. Altschul, *Economic Botany 22* (1968), p. 267.

Schultes, R. E. and R. F. Raffauf, *The Healing Forest.* Dioscorides Press, Portland, 1990.

Southon, I. W. and J. Buckingham, *Dictionary of Alkaloids.* Chapman and Hall, London, New York, 1989.

The Alkaloids (several eds.), Vols. 1-39, Academic Press, Orlando, New York, 1950-1990.

Webb, L. J., *An Australian Phytochemical Survey,* Bulletins 241 (1949), 268 (1952) C.S.I.R.O. Australia, Melbourne.

Willaman, J. J. and B. Schubert, *Alkaloid-Bearing Plants and Their Contained Alkaloids.* U.S.D.A./A.R.S. Tech. Bull. 1234. Washington, DC, 1961.

Willaman, J. J. and H. L. Li, Alkaloid-Bearing Plants and Their Contained Alkaloids. *Lloydia 33* (1970) p. 1.

Willis, J. C., *Dictionary of Flowering Plants and Ferns.* Student ed., Cambridge University Press, 1985.

Generic Index

T - #0178 - 101024 - C0 - 229/152/16 [18] - CB - 9781560228608 - Gloss Lamination